Guy Hull is a qualified and highly experienced dog behaviourist possessed of an encyclopaedic knowledge of dog breeds, cross-breeds and types, their histories and traits, as well as a passion for the historic and current role of the dog in Australian society. Based in New South Wales's Snowy Mountains, Guy has managed dog shelters and worked as a council ranger, holds dog training classes and provides dog behavioural consultations.

the Dogs that Made Australia

the Dogs that Made Australia

Guy Hull

HarperCollins*Publishers*

Aboriginal and Torres Strait Islander readers are warned that this book contains
names and images of people who have died.

HarperCollins_Publishers_

First published in Australia in 2018
by HarperCollins_Publishers_ Australia Pty Limited
ABN 36 009 913 517
harpercollins.com.au

HarperCollins_Publishers_
Level 13, 201 Elizabeth Street, Sydney NSW 2000, Australia
Unit D1, 63 Apollo Drive, Rosedale, Auckland 0632, New Zealand
A 53, Sector 57, Noida, UP, India
1 London Bridge Street, London, SE1 9GF, United Kingdom
Bay Adelaide Centre, East Tower, 22 Adelaide Street West, 41st floor, Ontario M5H 4E3, Canada
195 Broadway, New York NY 10007, USA

A catalogue record for this book is available
from the National Library of Australia

ISBN 978 1 4607 5645 4 (paperback)
ISBN 978 1 4607 1044 9 (ebook)

Cover design by Darren Holt, HarperCollins Design Studio
Front cover images: Armour [H20171] and Rifle [H2002.136] courtesy State Library of
Victoria; dog's head by shutterstock.com; dog's paws by Agency Animal Picture / Getty Images;
background painting courtesy State Library of New South Wales [972904]
Typeset in Bembo Std by Kirby Jones
Author photograph by Adam Klumper
Running greyhound footer by shutterstock.com
Printed and bound in Australia by McPherson's Printing Group
The papers used by HarperCollins in the manufacture of this book are a natural, recyclable product
made from wood grown in sustainable plantation forests. The fibre source and manufacturing
processes meet recognised international environmental standards, and carry certification.

For
Tony Parsons, OAM,
Bert Howard,
my sister, Vicki Hull,
and the dogs that busted themselves for Australia.

Contents

Author's Note

Read just about any account of the making of modern Australia, and you could be forgiven for thinking it was just the people who did all the heavy lifting. The true story of Australia's pioneering dogs has never been told, and what most Australians believe to be the origins of our working dogs ain't necessarily so.

I'd owned Australian cattle dogs for over thirty years and, as my understanding of the breed increased, I came to reject their accepted origin as a dingo x collie. There's plenty of dingo there all right, but the cattle dog has nothing of the collie about it. The more I looked at our cattle dog, and then kelpie origins, the more I realised to what extent we'd all been dudded.

Then about eight years ago, I stumbled upon an article with a startling, bleedin' obvious, and very plausible, explanation of the cattle dog's origins, and the seeds for this story were sown.

The author of that article, Albert (Bert) Howard, is a Second World War veteran who served in the Royal Australian Navy in the Pacific and was present at the Japanese surrender in 1945. In the 1980s he became heavily involved with the historical research for the privately published book *Over-Halling the Colony*, the story of the colonial beef cattle entrepreneur George Hall and his family. (Bert's late wife, Beryl, was George Hall's great-great-great-great-grand daughter.)

It was a Hall son, Thomas, who created the Hall's heeler, the forebear of our cattle dogs, and Bert's research included the

Hall's heeler as part of his decades-long family investigation.

Having sorted out the origins of the heelers, Bert turned his attention to our other great working breed wreathed in origin controversy, the kelpie. Noreen Clark's *A Dog Called Blue* and Tony Parsons's *The Kelpie* make use of Bert's cold-case heeler and kelpie research.

Thanks to Bert Howard the whole story of Australia's dogs can now be told. All the colonial heeler- and kelpie-related origin information herein, unless otherwise cited, is drawn from his *Australian Origins & Heritage Files*.

There is so little recorded about our early dogs (and nothing about the behaviour of prehistoric canines) that, where possible, I have elaborated on certain canine behaviours from a dog behaviourist's perspective – why wolves, dingoes or dogs acted in certain ways and what instinctive, human or environmental factors might have made them act so.

While I can manage the canine behaviour, it is beyond my powers to explain the behaviour of some of the humans instrumental in the founding of modern Australia. They have been roughed into the yarn with the measure of Australian levity that is their due. Some of the human cast even won supporting roles – well done to them – but when up against the dogs, most of the punters only auditioned well enough to land spots in the chorus line.

Our dogs are the real stars of this production, proving that eventually every dog, even a long-forgotten colonial dog, will have its day.

Introduction

Australia was all grown up by 1901. Few Australians were old enough to remember convict chain gangs and British redcoats, and fewer still thought the wintry, grey confines of the United Kingdom a healthier or better place to live. That year, the six colonies – New South Wales, Victoria, Tasmania, Queensland, South Australia and Western Australia – contrived to unite in nationhood. They had managed to gain independence from Britain without a single shot fired or a drop of blood spilled, gently wresting control of the asylum from the lunatics. It was very Australian in its doing.

The main driving forces behind Australia's rapid rise to maturity had been gold and wool. But wool had been doing it tough for a few years.

From 1896 to 1902, eastern Australia laboured under the thrall of El Niño's bone-dry misery, in what became known as the Federation Drought. It ravaged the continent with heatwaves, bushfires and dust storms, driving graziers to bankruptcy and decimating the nation's merino flocks.

Yet in 1898, at the height of the crisis, Australia's parched spirits had revived when a canine hero emerged, in an act that symbolised wool's defiance of the drought's devastation.

On Saturday 2 July of that dry, dreary year, a man named Jack Quinn walked onto the Royal Agricultural Society's

grounds at Moore Park in Sydney. He was there to take part in the sheepdog trial at the annual exhibition of the New South Wales Sheepbreeders' Association. It was the biggest event of its kind in Australia, and the sheepdog trial, known as the Sydney Trial, was the most prestigious event on the national trials calendar.

Today was the second round of the competition. Beside Jack Quinn was his blue kelpie, Coil. He and Coil had travelled from Cootamundra in midwest New South Wales to compete.

As they moved to their starting position, a shocked silence settled over the uneasy spectators.

Thirty-one of Australia's top working kelpies had run in the first round. Only eleven had progressed to the second and final run. In his first run, Coil had scored a perfect 100, and become the unbackable favourite to win the title. His faultless performance had been the first achieved in the three years of the Sydney Trials to that point.

Coil had all the right stuff: his mother, Gay, another Quinn kelpie, had won the inaugural event in 1896. Coil also had an inestimable advantage in his breeder and handler, the best kelpie man in Australia, and probably the greatest of all time.

Still, those advantages and a perfect first run counted for nothing, because Coil should not have been competing in the second round at all. In fact, he was lucky even to be alive.

Tragedy had struck on the evening of that historic first run, when Quinn and Coil were travelling to their Sydney lodgings. Full of himself after his heroics that day, Coil had prematurely leaped from the horse-drawn carriage as it neared their destination and had become entangled in one of the carriage's steel-rimmed wheels. His left foreleg was smashed.

Fortunately for Coil, Jack Quinn was studying veterinary science, and that night he reset the leg and dressed it in a cork splint.

The next morning, when any other dog would have been lying in pain and still in shock, Coil was hopping about so happily he convinced Quinn that he was fit to compete. After arriving at the grounds, Quinn consulted the sceptical officials and was reluctantly granted permission to give Coil the opportunity for his second run.

And what a run it was! Coil penned his sheep in just six minutes and twelve seconds, scoring another perfect round. *On three legs!*

It was an astounding feat that has become embedded in Australian working-dog folklore. The newspapers of the day christened him 'The Immortal Coil', and the name stuck. Coil was never to repeat that perfect score in Sydney, but his courageous heroics on three legs were more than enough to elevate him, and the kelpie breed in general, to legendary status.

* * *

The workaholic kelpie, Australia's home-grown collie, was the unsung hero of the success of Australia's wool industry. Coil's astonishing win exemplified the breed's stamina and unshakeable focus on the job.

It's incredible to think that not so long ago there were no dogs in Australia. Now we have one of the highest rates of dog ownership in the world.

Altogether, Australians own an estimated 4.8 million dogs – that's a dog for every five people. There are also many thousands of undeclared dogs living in backyards and working on pastoral concerns.

And Australia's rapid transformation from starving British penal colony to pastoral powerhouse has been due in no small part to the tireless exertions of the working dog.

This, then, is the astounding, untold story of the dogs that made Australia. It is the hidden story of the birth of agricultural Australia, as told from the dog's perspective.

And it has three main players: the people, the dogs, and its antithesis. While thousands of kelpies and German collies across the continent worked huge mobs of merinos with an untiring devotion to duty, the dog's antithesis was waging bloody and unrelenting war against those very same flocks.

So our story begins by exploring the origins of the world's most distinctive and influential 'wild dog'. It used to be *Canis familiaris*, like the rest of the dogs. It's now *Canis lupus dingo*: the dingo, Australia's wolf.

CHAPTER 1

The Dingo Conquers Prehistoric Australia

The dingo's domination of mainland Australia is a story of astonishing environmental conquest, punctuated by complicated relationships with Indigenous and European Australians. The dingo's story is melodrama on a grand scale, set against the backdrop of the most capricious, unforgiving settled continent on earth.

Ironically, if not for humans, there would be no dingo. It has often been tamed, but attempts at making a dog of it have been optimistic exercises in futility, because once the dingo outgrows youth's ambivalence it has no interest in being man's best friend.

The dingo polarises Australians. Some applaud it as an iconic native hero and others boo it as a shrewd, bloodthirsty villain. Yet the dingo is just a simple creature, naturally incapable of good or evil, a throwback to the age when the dog diverged from the wolf.

Australia's handsome little wolf was painted on rock and given mythical status by the Aboriginal people. More recently, though, it has been pigeonholed by medieval myth, old-world chauvinism and invincible stupidity. The new order declared war on the dingo 200 years ago because of its appetite for mutton and lamb. Relentless shooting, poisoning and trapping, and the longest fence ever built, purged and excluded the dingo from most sheep-raising districts, but the dingo wages war against wool still.

Declared enemy of the Australian wool industry the dingo may be, but dingo blood is present in some kelpie strains, and Australia's first native cattle dog, the Hall's heeler was derived from the dingo. Today the dingo remains a wild creature of the Australian bush, and the largest placental carnivore in the land of the marsupial.

Yet many Australians have no idea that the dingo, like most of us, is an import. That's hardly surprising. Only recently have advances in genetic science clarified the mystery of the dingo's origins, and the probable means of its arrival in Australia.

The tale of the dingo begins with the incredible story of the Eurasian wolf that became the dog that gave Neolithic man the big leg-up. Over thousands of years, in company with man, it traversed thousands of miles of land and sea and eventually reached prehistoric Australia. And there the dog abandoned its people and once again became a wolf.

* * *

Towards the end of the Ice Age, perhaps 12,000 to 15,000 years ago, environmental pressure caused some wolves to fundamentally change. One or more influences forced one or more Eurasian wolves somewhere in China to develop a dependence on humans and a tolerance of human proximity. Those wolves, possibly just a pair, began living like distant camp followers, existing around the fringes of a human group that was large enough to produce enough human waste, food scraps and excrement to sustain them.

The enmity between humans and wolves goes back a long, long way. These were the days when human and wolf hunting parties competed for game, to the point of driving each other off wounded prey or kills – particularly large animals that could not be carried home or, in the wolves' case, eaten on the spot.

These first human-tolerant wolves were probably young adults or juveniles not subject to the influences of an independent, human-wary pack. They may have been outcasts, perhaps sick, injured, emaciated or deformed, and it is highly likely they were smaller than average wolves and therefore less threatening in stature and behaviour.

It makes sense for the wolves that came in from the cold to have done so initially during the starvation months of winter. The smell of cooking food as well as the promise of warmth would have been hypnotic attractions. Perhaps they were unable to hunt to sustain themselves, or perhaps they found living near humans to be a better existence. The likely scenario is that a combination of all those factors, but primarily an easy living on human waste, caused that landmark behavioural change.

Prolonged, reasonably close association with humans would have started to desensitise the wolves, causing their instinctive fear to diminish somewhat. Over generations, a symbiotic relationship must have developed between the humans and their camp followers. Waste disposal would have been appreciated by the humans and a tolerance for the wolves would have developed among the people. Protection would have been a second benefit provided by them. As camp wolves followed the humans in their seasonal movements, the wolves' territory became fluid and based around the human settlement.

These wild, human-dependent camp wolves began to evolve into a different type of canine. Each successive generation would have developed a greater tolerance for, and dependence on, humans. Fear of humans in some individuals would have given way to shyness and in others to confidence, if not boldness.

In providing humans with waste disposal and protection, these wolves were displaying natural scavenging behaviours. But from the human point of view, they were earning their keep, acting as quasi-domestic animals. As time went by, they no longer behaved like typical wolves, but like semi-tame dogs: the product of human manipulation and a willingness of the wolves involved to modify their own behaviours.

The taming of the first wolf would have been organic and entirely voluntary. The first tamed wolf would not have been a captive. Captivity done right might encourage tameness, but it does not create domesticity. It would have been impossible back then to catch a wolf of any age and keep it in such a way that it was happy with the arrangement.

One day a camp wolf, probably a semi-tamed bitch, must have decided she had a better chance of survival by associating even more closely with humans. Her instinctive caution of humankind would have been overpowered by her desire to gain something she wanted. Food.

She had to have been starving. She would have understood that humans were a source of food. She would have voluntarily overcome her instinctive fear and approached a human, though still keeping a safe distance.

Maybe this hungry young wolf encountered a woman sympathetic to her situation. A woman's stature and voice are less intimidating than a man's. No doubt that little wolf bitch felt less vulnerable around this cave woman than she would have felt around a cave man.

The lady of the cave must have been used to dealing with the wild animals of her own species – her cave man and her cave children – and would have known just how to use food to motivate and train. Human habitation provides opportunities for many kinds of wild animals; perhaps she was already used to throwing scraps to the birds. She might have taken a maternal sort of interest in the lonely little wolf who sat at a safe distance when the men were away hunting and the camp was quiet, drawn by the overpowering scent of cooking meat.

Dogs are skilled in telling their owners when they are hungry. It is one of many juvenile wolf traits like tail-wagging and barking that have been encouraged and maintained in the domestic dog. The little wolf bitch would have made it known she was hungry by sitting in plain sight, alert but fidgety, whining, yelping and wagging her tail. At some stage, she must have plucked up the courage to come a little closer and take the food the woman threw for her.

Her approach would have been self-reinforcing. That first experience must have encouraged her to approach again for more scraps, coming even closer this time. More food would have been thrown, and the behaviour further reinforced, and so on.

It would have been a slow process of small steps, perhaps over a period of several months or longer. The wolf obviously understood that approaching the lady of the cave and sitting still worked to her advantage, but she would have always kept one eye on her escape route. She would never have learned to drop her instinctive guard.

So it may well have been a cave lady who established a basic rule of dog/wolf and human interaction: the one who controls the food has the power. Efficient training exploits a dog's desire to gain food. That technique started way back then. Granted, in training today some dogs prefer play opportunities or maybe a pat to a food reinforcement, but those dogs are invariably just too well fed and do not value food treats highly enough.

The lady of the cave would certainly have attributed human actions and reactions to the little she-wolf. In doing so, she would not have been too wide of the mark – whereas if we try that today we miss the target altogether. That's because over the thousands of years of the relationship, dogs' reliance on instinct hasn't changed, but ours has. We have abandoned much of that reliance because we've had dogs to do the relying for us.

The cave lady's aim, then, would have been to encourage the she-wolf to become human-like in her behaviour. Today, that would be a recipe for disaster, but human-like then was perhaps not so far removed from wolf-like.

Primitive people were patient. They had to be to successfully hunt, gather food, make clothing and tools, and generally survive in a harsh environment. Taming that first wolf would have been a painstaking process. There would have been plenty of prominent eyebrows raised – particularly those of the man of the cave – but there must have been enough positives to outweigh misgivings about encouraging the enemy into the fold.

So, how tame would that first wolf have become? The ice would have been broken, but she wouldn't have been capable of domestication. A wild-born she-wolf would never have lost her instinctive distrust of humans. Full human dependence would fall to one or more of her captive-born descendants.

There is a clear distinction between taming and domesticating an animal. *Taming* is the process by which an animal loses its fear of human contact. It is necessary before *domestication* – full human control – can occur. The taming of the first wolf turned her into a dog of sorts. A wild dog, certainly, but a dog nonetheless, and it was only a matter of time before full domestication would be achieved.

She would have mated with another human-dependent male outcast living on the fringes. Then, when heavy and gravid, she would have disappeared to whelp, and scraps would have been thrown to her partner, whose job it would have been to keep her fed. The people would have either taken a pup or puppies from her den, or appropriated them when she brought them near the camp once they were mobile.

The taming of the she-wolf might have raised eyebrows, but bringing the first wild-born puppy into the cave would have had an incalculable effect. As the puppy grew, it probably did not develop the wild creature's instinctive fear of humans. Instead, it became devoted to its people. That astonishing attachment would have been the first instance of faithful, human-focused, domestic-dog-like behaviour.

* * *

An international study of canine DNA conducted in 2009[1] concluded that this happened around 16,000 years ago. The world's first dog emerged from a Eurasian canid in central China that became extinct about 10,000 years ago.

The domestic Asian dog population would spread to all points of the compass over the next several thousand years, accompanying Neolithic people on their expansion beyond mainland Asia.

The dog, clever fellow, would have aided man in no small way as he started to capture and manage domestic stock, and establish farms and settled communities. The creation of the dog from the wild wolf was without question the greatest investment prehistoric man ever made. Without the dog's super powers, who knows how far we could have advanced by ourselves?

The dog played the role of kingmaker all around the world for about 10,000 years as Neolithic society rapidly developed. But kingmaker was a role the Asian dog could not play in prehistoric Australia.

* * *

The first people managed to navigate their way to Australia in rudimentary watercraft more than 60,000 years ago. In completing that remarkable journey, they traversed a 1000-kilometre-long archipelago and many straits, some spanning more than 70 kilometres.[2]

This was at least 45,000 years before the emergence of the Asian dog. And it would take another 10,000 years for that dog to reach Australia.

Making its way south through continental Asia and down the Malay Peninsula, in equatorial regions the Asian dog lost much of its luxurious coat. It must have been a useful all-rounder, herding water buffalo and wild cattle, and hunting wild boar, deer and babirusa, the unusual Asian deer-like pig. It would also have been an indispensable aid to the people colonising the Southeast Asian archipelagos. It appears to have been a companion on vessels of exploration, both as a guard and as a source of food. It obviously had a propensity for independence, and when old enough to fend for itself it was more than capable of making its own way in the world if it needed to.

Island-hopping in company with its migrating, colonising masters, the Asian dog travelled all the way to the Indonesian archipelago. And there, for thousands of years, it stayed, bottlenecked, until maritime technology and human endeavour could contrive to transport it to the last remaining frontiers: New Guinea and Australia.

This migratory bottleneck allowed the Asian dog to become fixed in type – becoming very similar in build and colouring.

There is a clear distinction between types and breeds. Greyhounds and whippets, for example, are separate breeds but similar types. Compared with the developing racy greyhound types of the Middle East, or the heavy mastiff types in Europe, by the time it reached Australia, the Asian dog had hardly changed since its development around 10,000 years earlier. In its purest surviving form – the dingo – it is virtually unchanged today. It is usually tan, black, or black and tan in colouring, has a coat of light to medium thickness and density, and is medium-sized, athletic and lean. Despite its malleable genetic make-up, the dingo has remained basic and unexaggerated – man's downsized wolf.

DNA genome sequencing studies conducted in 2009[3] indicate that the entire dingo population in Australia was founded by a small number of already domesticated dogs, possibly just one pair, perhaps 3500 to 5000 years ago.

By the time of the first dogs' arrival, more than 55,000 years after the first Aboriginal people reached Australia, sea levels had risen with the end of the Ice Age. But maritime technology must have developed sufficiently to accommodate an incredible number of explorations and migrations.

It appears to be a people now known as the Lapita who first brought the dog to Australia,[4] on one of their extraordinary voyages. Originating in southern China, they had ventured east to populate Polynesia. They shunned large land masses and only seemed to settle on small islands.[5] They were intrepid Neolithic explorers, and it is likely they knew Australia's northern shores well.

After a long sea voyage in very cramped conditions one can imagine the Asian dog puppies standing in the bows of the large Lapita canoes that closed in on a northern Australian beach, whining impatiently and barking at the smoky, exotic scents. Like family dogs with their heads stuck outside a station wagon window arriving at a beachside holiday destination, they were excited at the scent of promise.

The consternation of the Aboriginal people who met the first visitors on the beach that day is easy to picture. Perhaps they had never seen anything like the large sail-driven canoes, the strangers, or their dogs before. They would probably have been wary and prepared to defend their country, but the visitors were too few to have come with anything but peaceful intentions.

The Indigenous Australians would have been particularly intrigued by the visitors' dogs. They must have seen they had personality and a natural affinity with people, and that the strangers could communicate with them and control them. People and dogs have a mutually recognisable body language; it's the reason we got together in the first place. It is how we chiefly communicate with our dogs today, whether we realise it or not.

When the first dogs arrived, Australia's marsupial carnivores, the thylacine and the Tasmanian devil, still inhabited the mainland. The Aboriginal people probably thought the adult dogs were superficially like the thylacine, the tawny canine/feline-like marsupial carnivore with the striped back. The puppies must have seemed like noisy little black devils. Yet while the thylacine and the devil feared and avoided people, these dogs were bold; even the little puppies followed the strangers around. They played with children, had moods and squabbled. They swam in the sea and dug holes in the sand, and acted like part of the family.

The Lapita travellers would have been dependent on their dogs for survival when food was in short supply. Every boat would have carried several at least. But the best thing about the dogs – the Lapita may have communicated to the Indigenous peoples – was that they offered protection and drove off intruders.

The Aboriginal people would have valued the pottery that the visitors bartered in return for access to food and water. They must also have appreciated the two little tan puppies with the dark muzzles that the Lapita gave them. The terrestrial creatures of the Australian environment certainly did not.

* * *

The first Asian dogs weren't feral when they arrived here, that much is now known, but they were probably not what we would consider to be highly domesticated by today's 'fur baby' yardstick. They were independent types who could take or leave humans. Their previous owners hadn't given them the choice. Their new owners did.

Reliance on the dog's heightened senses and instincts through the millennia caused dog-addicted man to lose much of his own instinctive awareness. But the Aboriginal people, perfectly in tune with their environment and having managed just fine for around 60,000 years, did not need the dog's instincts to help them survive. These Indigenous Australians had no terrestrial carnivorous threats worth mentioning; instead, the greatest threat was posed by the harsh, unreliable environment itself.

Back when the first wolves decided to cast their lot with humans, both parties ratified an unspoken agreement. The wolf agreed to become the dog and help humans dominate nature. But in return, the people had to provide it with food, water, shelter, care, control and, most importantly, a job. The owners of those first two puppies, through the accident of their circumstances, would have been blissfully ignorant of any such requirements.

Dog ownership is a skill like any other. Unsuccessful dog ownership, as old a concept as domestication itself, is usually based on the irresistible appeal of small puppies and the failure of the owner to understand and deliver the necessities that develop canine dependence. Natural dog owners with obedient, highly socialised dogs make dog owning appear effortless, but gaining complete compliance from a dog is never as easy as it seems. Great dog people are born, not made.

The Aboriginal people's nomadic lifestyle could not provide the socialisation, restraint, and general management the domestic dog needs. Everyone loves a puppy, and they would have been no different, but once those two puppies had outgrown the cuteness of infancy they would have become, like any untrained dog, uncontrollable pains in the collective neck. After that, with no experience in dog management, the owners would have simply let them run at large. No control, no healthcare, no shelter, no food-tied dependence and no essential job. In doing so the contractual obligations were breached, the devil being in the fine print. The dogs, knowing their rights, quietly decided to activate the exit clause.

This would set in place a pattern of human–dingo relations that was still in place at the time of white settlement.

Even today, we must honour every facet of our side of the agreement, if our dogs are to reciprocate. It's the reason Fluffy stays with us. The other reason why today's properly managed, contented dog won't try to escape its suburban yard is simply because it doesn't need to, and doesn't know it can. Everything it needs is at home, and because it is desexed that instinctive urge does not tug at it to escape and sow its wild oats.

The dog of today is not so different from the Southeast Asian dogs that arrived here thousands of years ago: break the contract, deprive it of the essentials, and if it knows it needs to support itself, it will. And that's just what the dingo's ancestors in northern Australia did.

The desertion of the first dogs would have been a process of degree, and not at all gut-wrenching for the Aboriginal people, as it would be for us to come home to find the side gate open and Fluffy gone. Those first dogs would have needed to find much of their own food by scavenging, stealing or hunting, and they would have started young. Small animals would have been easy prey, and as the dogs became more proficient hunters with age, they would have become more independent and strayed further and further from camp. Feral dogs in the making.

Within a year or so they would have been mating, and soon the very first litter of puppies would have appeared – puppies

born with a tolerance for people, but the bush in their blood. A temporary pet at best.

And so the feral dog population exploded. Every new generation would have been born a degree wilder, until its type became fixed as the canine we call the dingo: the Asian dog, adapted to the tough Australian environment.

* * *

The eastern Top End of Australia is a steamy land of thin, leached red soil, massive termite mounds, straggly eucalyptus and melaleuca, tall cane grasses, and belts of rainforest along the watercourses. It was regularly burned, firestick-farmed, by the Aboriginal people, providing them with easier access to wallabies and a variety of smaller marsupials, goannas, magpie geese, insects and other terrestrial game. There is an abundance of fish and crustaceans in the clear, freshwater streams. On the coast, the estuaries teem with birds, mud crabs and fish, and the beaches are a constant source of forage.

As the dog that became the dingo deserted its first owners, it made its way alone in an environment that had plenty to offer. But there was plenty of danger too. Venomous snakes, large pythons, and crocodiles, which display a strong preference for dogs, would have accounted for the unwary. It was the perfect environment to trim the dross off the expanding gene pool.

Radiating out from the point of their original liberation, the new dingoes colonised the tropical coastlines eastwards and westwards, then southwards, territory by territory. Within a short period of time, they dominated every habitable and barely habitable environment in mainland Australia, as well as some nearby islands to which they were probably taken by

Aboriginal people as small puppies. The invader swept through the continent with astonishing speed because there was little to slow its progress. Aiding that progress would have been the trading of puppies between Indigenous people.

It may have taken the dingo less than 100 years to infiltrate the entire land mass. That estimate is based on the fact it took the European red fox just eighty years to chew its way through the native fauna on its way to the Kimberley in Western Australia, after being released by an upper-class fox-hunting numbskull in Victoria in the mid-nineteenth century.

* * *

The dingo's carnivorous marsupial competitors had no answer for its guile or its pack-driven organisation and aggression. The dingo has long been blamed for the mainland extinctions of the thylacine and the devil, and that's fair enough; he's always had guilt written all over him. Lately, though, science has suggested the dingo may only have *contributed* to the thylacine and devil extinctions, alongside climate change, an increase in the Indigenous population and advancements in their hunting weapons.[6]

Those changes might have moved the goalposts for the mainland thylacine and devil, but in the rule book of apex canine predation, the local heavies have always taken out the competition. There are no wild dogs anywhere in the world living near populations of old-world wolves. The wolves, having got their hands on a copy of the human–dog contract, have never forgiven the dogs' treachery, and kill them on sight. The development of the wolfhound – bred to run down and kill its wolf ancestors – is still quite a sore point. Wolves do not

tolerate the smaller canids either. Coyote? Jackal? Fox? You wouldn't want to be one of *those* lightweights, lured by the hypnotic scent of a kill, only to find the local wolf pack tucking into the main course. You'd be dessert!

Dingoes are known to actively prey on foxes and cats, but eliminating the opposition isn't limited to the canids. The big cats kill the not-so-big cats, and the not-so-big cats kill the even smaller cats. Lion males kill other males' cubs; lions and leopards kill African hunting dogs whenever they get the opportunity. Apex predators hate everyone, so you wouldn't imagine the dingo and the thylacine ever spent much time swapping notes on the local prey species.

The dingo dominated its new environment because it was more aggressive than the local native species, it had a gang mentality and, in many cases, was larger. As an introduced predator, it also enjoyed an unfair advantage, as native prey species had not developed defence mechanisms to help them avoid, or survive, predation.

It is inconceivable that the thylacine could have coexisted with its aggressive, highly mobile and organised new neighbours. Someone had to give ground, and it wasn't going to be the mob. The dingoes would have killed the thylacines, or harried them off kills and pushed them out of their home ranges, and for a territorial predator, that always ends up ugly. The devils, being small, nocturnal and extremely noisy when feeding, would have been easy targets for these opportunistic invaders too. It was only in the safe haven of Tasmania, which the dingo could not reach, that the thylacine and the devil would remain unmolested for the next 5000 years.

* * *

This 1888 woodcut by Samuel Calvert depicts three dingoes watching grazing kangaroos in an old-world forest setting. The dingo's reputation for cunning was as widespread as its reputation for savagery. (Courtesy of State Library of Victoria, IAN26/05/88/92)

The dingo's wolf-like and human-like social pack structure also helped its rapid spread and survival. Dingoes pair for life and are family-oriented and devoted partners and parents.

The domestic dog isn't. Human meddling has made the post-mating role of the domestic sire redundant, and the bitch as a caregiver mostly redundant after a month. The domestic bitch usually has two seasons (heats) a year, unlike the wolf and the dingo, which have just one. The domestic male, the 'dog', is a promiscuous fellow, and will mate with as many bitches as he is able. Once he mates with a bitch he takes his ease and has nothing more to do with her or the puppies. The bitch nurses her puppies to around four weeks, alone, before her owners begin the weaning process, and

gradually over the following month she has less and less to do with them.

For the dingo, it's family versus everyone and everything. The typical dingo pack consists of a bonded pair and their offspring, and the odd adoptee, which could be of almost any age. In the bush, the dingo male and sometimes some of his older daughters help his partner in raising the puppies and share the workload of keeping the nursing bitch and the puppies of weaning age fed. Dingo puppies are nursed much longer than domestic puppies, up to ten weeks – and with the emergence of their needle-sharp milk teeth, that is a labour of love for mum! As dingo puppies mature they stay in the family pack.

And in years gone by, some of those puppies were appropriated by the Aboriginal people well before weaning and spent the first year of their lives as tamed, but never domesticated, pets. This probably started with the offspring of the first pair of Asian dogs that took to the wild. Yet the popularly accepted view of the dingo as a faithful canine companion in the traditional Aboriginal social context is a mix of romanticism, ignorance and the inability of some observers to differentiate between a dingo and a domestic dog.

* * *

Dingo puppies are certainly as appealing as any domestic breed. Sourcing puppies would have been simple for a people who had an intimate knowledge of their country. They knew the local dingo packs, where their dens were located and which dingo bitches used them.

Dependent infant puppies were easy to manage and good for a cuddle, stayed close by, ate little, and were playful and entertaining. The Indigenous Australians loved their young dingoes. Childless women, particularly the elderly, were usually the owners of the largest number of dingoes, sometimes owning several.[7]

The people, observant and respectful of all the animals of the bush, would have noticed that the dingoes, in many ways, were just like themselves. They lived in family groups in which the oldest male was the leader. They cared for each other and acted cooperatively for the group's benefit. They loved their young, and like the people, they sang together at night. The dingo might have been a latecomer, but it provided a type of companionship the nomadic Aboriginal people could relate to.

For a people living hard, the addition of a flamboyant, child-like pet would have provided some entertainment – for a while. A temporary pet was all a dingo ever amounted to, and that was all the Aboriginal people expected it to be.

Moreover, contrary to common perception and even observations made by early colonists and others a century later, the dingo could never have made the grade as a large game-hunting companion for Indigenous men. Cooperative hunting dogs are the result of hundreds of years of selective breeding for specialised work. Obedience and a propensity to work in cooperation with humans, the absolute requirements for a hunting dog, have never been the dingo's strong suit.

Untrained, uncontrolled adolescent dingo pups out on a serious hunt with Aboriginal men is a ludicrous concept. They'd get a nostril full of kangaroo scent and take matters into their own hands – and would be dodging nulla-nullas if they were silly enough to come back to camp when the game was given up for lost.

No, the dingo would have been a worse-than-useless hunting companion for a people whose principal hunting strategy was stealth. Young pet dingoes would have been of some use in helping the men practise their hunting skills, in detecting the presence of smaller game in hollows, up trees or down burrows, or in harrying slower game like possums and reptiles. But no Indigenous families would have been putting their hard-earned on their dingo pups to get them through the lean times.

Early reports by missionaries and scientists of 'domestic dingoes' hunting in cooperation with Aboriginal men were obviously well-intended, but they have caused a great deal of overestimation of the dingo's capability and role in traditional Indigenous society.

It has so far proven impossible to make a human-focused domestic dog of the dingo. The independent, uncooperative nature of the feline is well known and, as every surviving lion-tamer will attest, the big cats are certainly somewhat trainable as performing animals. However, captivity-bred wild canines – wolves, dingoes, and the like – are not.

It appears certain that early observers saw dogs of a distinctly dingo-like appearance join the hunt and called them for dingoes. But these were surely camp dogs, part-bred dingo-domestic-dog hybrids capable of higher standards of training and cooperation than the dingo. Indigenous people were quick to take advantage of the cooperation of domestic dogs and proved to be very capable dog handlers. Domestic dogs crossbreeding with dingoes was inevitable, and that early progeny is always very dingo-like in appearance, while probably retaining more of the domestic dog's tractability.

There is a world of difference between tameness and domestication. There are always exceptions and certainly some people – Indigenous and non-Indigenous – have raised some tamed dingoes to almost domestic-like standards, but the fruits of those labours have never approached a lasting, generational form of domestication.

Another widespread belief is that dingoes paid their way as hot-water bottles on cold nights. Little dependent puppies seeking warmth and company, maybe. But maturing dingoes were not usually available, or willing, to volunteer for those duties.

Dogs were. As we'll see, as soon as European domestic dogs became available, they quickly superseded the dingo as the Aboriginal peoples' canine companion of choice. Snuggling up to people is very much a dog thing. Winter nights in subtropical, temperate and alpine Australia often plunge to sub-zero temperatures. The Aboriginals would have welcomed the extra warmth of their dogs lying close by them, providing they didn't fire-hog.

The further claim that dingoes acted as camp guards is a clear case of mistaken identity. There is an obvious distinction between a watch dog and a guard dog: a watch dog warns, and a guard dog protects. Any dingo at camp would have growled at the approach of a stranger, or a bird flapping in a tree at night, but growling at the wind hardly makes it a camp guard.

Australian fauna work night shifts, so naturally the wild dingo is most active from dusk through to dawn. When their human families were ready to snuggle up under skins for the night, the camp dingoes would have been starting to wake up, particularly when their wild kin cranked up the howling – and there would have been plenty of that back in the day. A hungry dingo is an active dingo, and unless the camp dingoes were extremely well fed – and usually they weren't – they would have been hungry and restless, trotting off into the scrub for the night shift, the dog watch.

* * *

Once it outgrows the playful indolence of childhood, the dingo, unlike the domestic dog, begins to think of matters far weightier than playing with the kids. For the maturing pet dingo, it is all about survival and procreation, and the call of the wild always overpowers any human attachment. The people knew that as soon as their dingoes reached sexual maturity, they would be gone.

Dingoes mature more quickly than all domestic breeds. This is commensurate with their shorter life expectancy, five to six years being a fair old age for a wild dingo. Dingo pups would have neared sexual maturity at around twelve months of age, and camp life with them would have become testing.

Any untrained, poorly socialised adolescent dog is a pushy pain, and hungry young dingoes would have been seriously competing for food they had made no contribution to collecting. Imagine trying to manage a barbie on the ground with a dozen half-starved, untrained, unrestrained, ill-disciplined, teenaged dingoes hovering about!

The dingo is notoriously sensitive and thin-skinned, and takes any chastisement very much to heart. As the camp dingo gained sexual maturity it would have found itself an ostracised, scavenging fringe-dweller until the call of the wild convinced it that it'd be better off making tracks. As youngsters matured they either made off of their own volition, or became so troublesome they were driven off or, if necessary, knocked on the head. And once they handed in their notice, they had nothing more to say to their former masters.

·We can only speculate on how most dingoes fared when they pulled the pin on camp life and returned to the bush. Many of them would have been in poor condition through malnutrition and parasite infestation, and critically short on dingo life skills. Yet finding a feed in the wild would have been easy enough, even for a solitary dingo. Having followed Aboriginal women about foraging, they would have learned plenty, and with their eclectic tastes, most outcasts would have fared better than they did in camp.

Any dog, dingo or wolf deprived of company for a prolonged period is a miserable creature. The solitary life is not for any of the large social canines. Instinct would have driven outcasts to seek the company of their fellows.

Inappropriate behaviours learned in camp, like queue-jumping or snatching food, would not have been tolerated in the wild pack, but recruited camp outcasts might have brought new hunting skills learned from their time with people. They also would have infused the gene pool with the apparent tolerance of loose coexistence with Aboriginal people that dingoes are known for.

Not all camp outcasts would have found acceptance in the wild. There would have been instances where a fresh crop of excommunicated yearlings formed the nucleus of a new fringe-dwelling pack, or added to an extant one. Rejected by both their former owners and the wild packs that owned the surrounding territory, these dingoes would have been trapped in a dingo no-man's-land. They may have been displaced or traded from another region, but it is highly likely these types were timid, lame or those with a warmer disposition towards humans. Following white settlement, the neediest ones might have been among the first to crossbreed with domestic dogs, which would eventually replace the dingo as the canine companion of choice in Indigenous communities.

* * *

The phenomenal success of the dingo in Australia vindicated the Asian dog's desertion of the campfire. The dingo possessed the physical and instinctive wherewithal to adapt to and quickly dominate the entire Australian mainland. It shared the role of apex predator with the Indigenous people, and the natural biological balance was maintained.

Unlike the persecuted, demonised wolf of the northern hemisphere, the dingo was appreciated by the equally unique people of Australia. Neither the dingo nor its former people harboured any hard feelings over not being able to get together permanently, but lived and let live. Easy come, easy go. No wolf had ever been given such an opportunity.

But the world was shrinking, and time was coming to the hitherto timeless land. The jostling imperial powers of Europe had developed an insatiable appetite for new colonial conquests. A hard, foreign people were coming, and with them an intolerant old-world order, and the unique coexistence between the dingo and the first people of Australia would soon be smashed under the hammer blow of British colonialism.

CHAPTER 2

The British and the Dingo Get Acquainted

It was 1699, and William Dampier – onetime privateer, global circumnavigator and scourge of the Spanish – was heading for the east coast of New Holland, as Australia was then known, when he found himself along the uninviting coast of northwest Western Australia. And he certainly didn't find the local dingo to be an impressive creature.

'There are but few land animals,' he later wrote. 'I saw some lizards; and my men saw two or three beasts like hungry wolves, lean like so many skeletons, being nothing but skin and bones.'[1]

If the local wolves were starving, there clearly wasn't much to entice foreigners. Dampier continued northwards in his quest for the east coast, misfortune ensuring that he never made landfall in Australia again.

The Dutch had known of New Holland since Willem Janszoon dodged a continual barrage of spears from the folk of the western side of Cape York on his seemed-like-a-good-idea-at-the-time visit in 1606. He made no mention of wolves or dogs. Neither did Dirk Hartog, who landed on an island off Shark Bay in northern Western Australia in 1616. Underwhelmed by his discovery, he left a message hastily scratched on a pewter plate, nailed it to a post and made off with all haste for Batavia to the north, his original destination.

It would be seventy-one years after Dampier's visit before the British laid eyes on the little ginger wolf again. Lieutenant James Cook of the Royal Navy had been given command of the bark *Endeavour* and dispatched with a party of scientists to observe the transit of Venus across the sun in Tahiti. The Admiralty also gave Cook a nudge and a wink and told him to be a good fellow and find the great southern land while he was down there. Ever resourceful and devoted to duty, Cook fulfilled his every instruction and found the east coast of New Holland, stepping ashore at what he initially called Stingray Bay, which at botanist Joseph Banks's urging was renamed Botany Bay. Cook and his landing party received a frosty reception from two Aboriginal men, and an exchange of musket fire, stones and spears resulted in one of the locals being wounded. The local people then made themselves scarce. Undaunted, Cook's shore party took the opportunity to stretch their legs and glimpsed their first marsupials and their first dingo paw prints, as the captain recorded:

> In the woods between the Trees Dr Solander had a bare
> sight of a Small Animal something like a Rabbit, and we
> found the Dung of an Animal which must feed upon
> Grass, and which, we judge, could not be less than a Deer;
> and the footsteps of another, which was clawed like a dog,
> and seemed to be about as big as a wolf.[2]

Cook had the Union Jack planted and the basic details of their visit recorded on some trees near the place where they replenished their water. These things done, and with the anchor aweigh, Cook duly sailed north, mapping the eastern coastline, until *Endeavour* had a prang with the Great Barrier Reef off what is now Cooktown in far north Queensland.

Careening *Endeavour* on the southern bank of the aptly named Endeavour River, the crew effected the lengthy repairs under the instructions of the ship's carpenter, while Joseph Banks, in company with his scientific colleagues, went off collecting flora and fauna specimens.

Despite the best efforts of the scientific men, it was an ordinary crew member who was the first to spot a dingo. He took a pot shot at it but missed. Banks noted in his diary on 29 June 1770 that the midshipman involved was an American, who said the dingo looked exactly like a North American wolf.[3]

Exactly like an American wolf? Perhaps the midshipman didn't know what a real wolf looked like. Or was it possible that he was familiar with the obscure Carolina dog, North America's square-peg-in-the-round-hole wild dog, which *does* bear a remarkable likeness to the dingo? We will never know.

What we do know is that the young fellow's actions and attitude towards the dingo were precursors of the prevailing British attitude towards anything that resembled the wolf they had driven into extinction in the sixteenth century. Lieutenant Cook recorded much the same entry in the ship's log.

A little over a week later, Tupia, Cook's interpreter from the Society Islands, saw another dingo, but for all their searching it was the *dingo* that came to *them*. The captain's last entry relevant to the fauna of New South Wales states that he had seen a dingo that frequently scavenged about their camp for discarded bones.[4]

Endeavour was eventually repaired and her crew gingerly sailed her out through the maze of the Great Barrier Reef, leaving New South Wales without close examination of the little wolf, but Cook did have the presence of mind to stop at a small island at the tip of Cape York (Possession Island) and take possession of New South Wales – roughly the eastern half of the continent.

Australia had a mere eighteen years' more breathing space before the wolf-allergic British would return to put up the 'Under New Management' sign, and dump their human dregs on New South Wales. An indiscriminate people were coming, a people who, with help from their domestic dogs, would destroy the dingo's gains of 5000 years and up-end the natural order.

Aggressive people own aggressive dogs, and there were few more aggressive than the empire-building British. Nothing in Australia would ever be the same again.

* * *

At this point it's useful to take a look at the dog's various societal and utility roles in Britain at the time when convict transportation had begun and the first British colonists arrived in Australia.

Dogs have diversified into many types and breeds since their domestication, and the British Isles have created over eighty distinct modern breeds. France, a much larger country, has created sixty-eight breeds, and Germany, also larger, forty-nine. While various breeds and types have come and gone according to people's needs and whims, there has never been a nation to equal the British in their love of breeding dogs.

Dr Johannes Caius (Dr John Keys), physician to Queen Elizabeth I, wrote the first English book of dog breeds, *Of*

Englishe Dogges, in 1570 (published in Latin then translated in 1576).[5] Dr Caius mentions many breeds that are still around today, and classifies them by means of quaint descriptions.

It is apparent that by the sixteenth century, dogs the world over had evolved into seven main 'types': the **mastiffs**, the **sight hounds**, the **scent hounds**, the **gun dogs**, the **pastorals**, the **spitzes** and the **terriers**. While many of today's breeds have emerged within the last 150 years, all of them have descended from one or more of these basic types.

Since the establishment of the feudal system, hunting had been a matter of sport for the well-to-do. Initially, hounds, strong-willed, swift dogs that detected game through eyesight or scent and chased and killed their quarry independently of their handlers, were the only form of hunting dog. But in the sixteenth century 'fowling pieces' first appeared in Europe – smooth-bored, shotgun-type firearms that emitted a cluster of small spreading shot – and gun dogs began to evolve from hounds. They were used to hunt smaller game like rabbit, hare and game birds and worked in close cooperation with the hunter. It was only the upper classes who had access to firearms and could maintain dogs for leisurely pursuits, while subsistence hunting remained a necessity for the rural lower classes.

By the late eighteenth century, rural peasants in Britain still relied on coursing dogs – those that detect game by sight and run it down – in order to catch meat. Types such as the lurcher offered them a chance of securing the odd humble rabbit, and were affordable enough to maintain, though merely keeping a lurcher brought the owner under constant suspicion of poaching.

The lurcher was, and still is, a type rather than a set breed. Its ability to do the job was always more important than what it looked like, so there was some variation between lurcher types. The most popular was the greyhound x collie. The aim was to produce a running hound that retained the working dog's obedience and docility. They were rough-coated dogs, trained to silently disappear if they heard, saw or scented anyone other than their master because poachers were essentially thieves trespassing on Crown land or private estates.

Meanwhile, on his rambling estate, His Lordship kept a wide variety of breeds. There was a pack of foxhounds for organised hunts, to run down and devour every last piece of the unfortunate fox. His harriers and beagles pursued hare and rabbit. He also owned an array of little smooth-coated, earth-digging terriers (*terra* is Latin for 'earth'), who went to ground hunting foxes and badgers, native animals considered vermin by His Lordship.

The otter was a truly despicable creature that had the criminal temerity to catch and eat His Lordship's trout and grayling, damn it, so His Lordship kept a pack of otterhounds: rough-coated hunters, part terrier, part hound, famed for being able to scent above water. What true sportsman *wouldn't* keep such animals? For there would be no casting a fly to a brown trout unless one's otterhounds remained vigilant and swept the streams clear of the otters.

The waterfowl that frequented His Lordship's waterways were, of course, his property. There was no greater comfort than a steady retriever that could brave the iciest waters, to deliver downed ducks to His Lordship's hand. And a stroll through his rolling downs or along his hedgerows with his fowling piece was an intolerable inconvenience without a

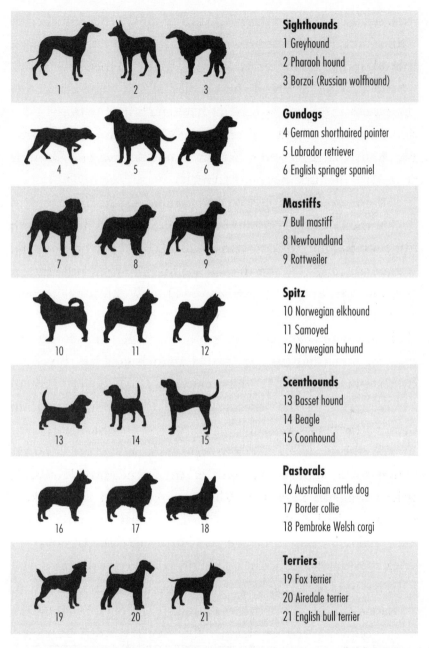

Sighthounds
1 Greyhound
2 Pharaoh hound
3 Borzoi (Russian wolfhound)

Gundogs
4 German shorthaired pointer
5 Labrador retriever
6 English springer spaniel

Mastiffs
7 Bull mastiff
8 Newfoundland
9 Rottweiler

Spitz
10 Norwegian elkhound
11 Samoyed
12 Norwegian buhund

Scenthounds
13 Basset hound
14 Beagle
15 Coonhound

Pastorals
16 Australian cattle dog
17 Border collie
18 Pembroke Welsh corgi

Terriers
19 Fox terrier
20 Airedale terrier
21 English bull terrier

The seven general types of dog in modern-day representations. Sight hounds and spitzes are the most ancient and would have changed very little over the last several hundred years, with pharaohs and greyhounds unchanged for thousands of years. Some of the mastiffs are the next longest types that have remained unchanged; the three represented here are relatively newer breeds.

rock-steady pointer to show him where a pheasant crouched in hiding, or without a steadfast setter that would set (sit) when it had scented anything worth blowing to smithereens.

But nothing warmed the cockles of His Lordship's heart more than the good fortune of having a trusty spaniel by his side to retrieve the brace of woodcock he blasted out of the sky. And what a blessed relief it was not to have to bend over to pick up the little blighters himself!

Of course, His Lordship had far more pressing issues to deal with than simply tramping about his estate, shooting with his peers, quaffing brandy and complaining about his gout and wicked tenants. Holding the seat for his borough in the House of Lords, His Lordship often travelled to Westminster to vote on weighty matters of State.

In the late eighteenth century, his hobby horse was the growing fractiousness of the commoners. That his very constituents used their unemployment and poverty as an excuse to trespass and pilfer from his estate infuriated him, but there's your starving commoner for you.

His Lordship's journeys to London were no less encumbered by the presence of these commoners. While approaching the hazy city in his carriage, he often found his progress slowed by drovers taking large herds of cattle to the Smithfield markets. Driving the cattle were the pastoral (or stock-working) dogs, either bobtails, collies, or drover's curs – a leggier bobtailed dog of fearsome reputation that would have a profound influence on the beef industry of New South Wales in the following century. Until 1796, working stock dogs were exempt from dog taxes, provided their tails were docked.

Fortunately, while His Lordship was away at Parliament, bellowing across the floor of the Upper House at Whiggish intrigues – he hated a Whig – he had able staff to manage his estate. He retained kennel masters to maintain his dogs and hounds, and gamekeepers to guard against the poaching and hayrick-burning of his ungrateful tenants. Just as most of his leisurely diversions were connected with sporting dogs, so aggressive man-catching breeds also maintained the protection of his sporting preserves. His Lordship's estate and its rich diversity of game were too great a temptation for his starving tenants. They wouldn't scruple to hand-slip their lurchers to silently run down His Lordship's rabbits of a night, or during inclement days when the fog or rain dampened their scent. So His Lordship maintained English mastiffs and bull mastiffs to guard against local poachers.

* * *

Until recently, crimes such as nicking His Lordship's rabbits had been a felony punishable by hanging outside Newgate Prison. But now, thanks to the mercy of His Lordship's Tory majority, and the small matter of the deafening public outcry against ghastly executions and atrocious, overcrowded prison conditions, things were different. The local poachers subdued by the mastiffs and taken before the local magistrate at the assizes – usually His Lordship himself – could look forward instead to a lengthy stint in a prison hulk on the Thames. And soon the wretches would pick up their chains and shuffle off in their irons to join the first fleet of convicts transported to serve their sentences in Botany Bay in New South Wales. A jolly good thing too, and good riddance to bad rubbish, thought His Lordship.

In many ways, life for the lower classes was even harder in London than on His Lordship's estate. The London of the time

was stinking, smoky and grotesque, still sobering up from the mass gin addiction of a few decades before. Unemployment was rife, and just like the country poachers, the starving resorted to petty and not-so-petty crime to survive.

The common man, uneducated and often unemployed, had few wholesome diversions. England had acquired an unquenchable thirst for blood sports during the Roman occupation. Barbaric bear-baiting and bull-baiting spectacles were still popular, and legal, in the late eighteenth century.

London and other cities suffered constant vermin problems, and there was always an overabundance of rats. Terriers took part in rat-killing competitions to entertain the punters in drinking houses throughout Britain. By the late eighteenth century, England had developed an astonishing twenty-two terrier breeds, many of them smallish and agile, and hard-wired to destroy small animals. Manchester and fox terriers were the favoured ratting breeds and were the common companions of the types of common person who often engaged in less-than-savoury activities on London's streets, sometimes just to survive.

But crime was crime, and nabbed by the thief-takers or the Bow Street Runners (London's first police), these unfortunates also found themselves shuffling in clinking ankle chains aboard the convict transports of the First Fleet, banished to a world they could not comprehend.

Britain was always at war with someone, and the powerful upper classes were also constantly at war with the powerless lower classes. They had established a class segregation that became so vehement that they gaoled en masse the people they found most repugnant. And when the gaols were full, they shipped them off to the other side of the world for the most trivial crimes.

* * *

The eleven ships of the First Fleet sailed from Spithead and Portsmouth on 13 May 1787, bound for New South Wales. The livestock carried on board were two stallions, four mares, one bull, four cows, one bull calf, forty-four sheep, nineteen goats and thirty-two pigs. The poultry on board were 122 fowls, eighty-seven chickens, thirty-five ducks, eighteen turkeys, twenty-nine geese and a rabbit. The dogs listed were assorted puppies, and Governor Phillip's greyhounds. It is apparent, though, that other dogs sailed on the First Fleet: journal entries by various writers refer to spaniels, terriers, 'wiry' greyhounds, and Hector, a Newfoundland. But apart from Hector and the greyhounds, the dogs' owners, breeds and how many of them there were is not known. It is almost certain they would all have belonged to marine or naval officers.

Not every dog taken onto the ships of the First Fleet would make it to New South Wales. The first mention of a dog on board was an unhappy one. Marine Lieutenant Ralph Clark, serving on the transport ship *Friendship*, noted with sadness and anger in his journal that his puppy, Efford, had been lost overboard.[6] He strongly suspected the first mate, and would have had the men give him a good thrashing if he could have proved it. Lieutenant Clark had owned the puppy for less than a month, and it seems poor little Efford and the first mate were not on good terms. Perhaps the first mate was a taut fellow who ran a tight ship and objected to the mess an uncontrolled puppy inevitably makes.

But little Efford was not the only dog lost overboard on the journey. Marine Captain James Meredith suspected his dog,

Shot, had got the old heave-ho, and the records show that five dogs in all were lost in similar circumstances.[7]

The many dogs that did make it to New South Wales would profoundly influence the fortunes of the fledgling colony.

* * *

The First Fleet arrived in the harbour Cook had named Port Jackson on 26 January 1788, after Governor Phillip rejected Botany Bay as an unsuitable site.

The local Cadigal, a prominent clan of the Eora people, called the harbour Cadi. It was a pristine environment in which the eucalyptus forest grew to the very edges of the shore. In the height of summer, it was a sweltering place, full of the noise of cicadas and birds.

The alien stench, let alone the sights and sounds, of over 1300 people, 100 mammals and nearly 300 poultry, cooped up in the eleven ships of the First Fleet for eight months, would have been overpowering for the Eora, and irresistible to the dingoes. The breeze would have wafted the nautical cries of the seamen and the reek of the passengers up the harbour ahead of the slow-moving flotilla. The dingoes' noses must have been pointing up into the breeze, scenting the ships of the fleet, well before they saw them.

The whiff would have increased as the convicts disembarked, and for the dingoes it must have seemed like the scent of opportunity. The irresistible smells of food, cooking, humans and their waste, animals and their waste – everything – would have interested both the Eora's pet dingoes and the wild pack.

As soon as the dogs of the First Fleet came ashore they would have run around smelling the wee-mail on the trees, and they would have cocked their legs over every existing scent-marker they found. The dingo may technically be a wolf, but it has none of the wolf's enmity towards the domestic dog. Both dogs and dingoes would eventually consort, and the dilution of the pure dingo on the eastern seaboard had begun.

We have all been mystified, bemused or horrified by our dogs' attraction to stinking things. They love rolling in the aroma of cow manure or any other putrefying matter. The more nausea-inducing, the more they love it.

We humans are visual creatures. We are taller than our dogs and can see much further than they can. Spend a little time moving about on your hands and knees and you'll quickly appreciate why dogs have such a different view of the world, and why their primary sense is smell.

We have around 5 million olfactory receptors in our noses; it sounds like a lot, but the dingo has around 200 million. Dingoes, dogs and wolves 'see' the world – a world we cannot conceive – through their noses. Scents that are often undetectable to humans tell them everything about their environment. They then use their eyes or their ears to locate and confirm what they've already scented.

Dogs are not so good at identifying stationary objects, but can detect the slightest movement. Many animals use camouflage and stillness as their prime methods of avoiding detection, but hunting via scent is much more effective than hunting by sight.

All living things exude a scent. Scent given off by quarry remains detectable for varying lengths of time, depending on the type of scent, weather conditions and terrain and the

scenting ability of the dog. It doesn't matter how still and well hidden a prey animal is; if there is a movement of air and a predatory canine is downwind of it, scent will betray prey every time.

Most domestic dogs use only one primary sense for hunting: either scent (bloodhounds, bassets, beagles, foxhounds), or sight (greyhounds, borzoi, deerhounds). Scent hounds are persistent and methodical in pursuit; sight hounds pursue game they have seen and are reliant on speed for success.

A few sight-hound breeds, such as the Maltese kelb tal fenek, or pharaoh hound, the Sicilian cirnecco dell'Etna, the Spanish podengo and the Ibizan hound, use hearing, scent *and* sight to hunt. These primitive dogs have large, erect ears and are the exceptions to the rule. The dingo, like the wolf, the coyote and the jackal, also uses all three senses; survival necessitates it.

Dingoes, dogs and wolves use scent in many other ways too, such as territory-marking and mutual identification. There is more information exchanged through leg-cocking, bum-sniffing and cow-poo-rolling than we will ever understand. Even modern dog breeds that bear no resemblance whatsoever to the wolf are motivated by primitive – and to us, at times, revolting – habits. The modern dog's appearance may be dramatically altered, but deep inside, the wolf remains.

There would have been the same bum-sniffing, stiff-legged sizing-up and squabbling as the dogs from the different ships of the First Fleet came ashore and mixed. Still, things would eventually have settled down as a hierarchy was established, because dogs are sociable creatures and instinctively need the company of their own kind. It would take a while before the local tamed and wild dingoes and the dogs of the First Fleet met. It wouldn't happen overnight, but it would happen.

And so formed New South Wales's first domestic dog pack – the first of many for the infant colony – and that pack and their offspring would be making up their own rules before long. They had to, otherwise, like their owners, they would have faced starvation.

* * *

The early writings of the First Fleeters hardly mention the colonists' dogs. That's not surprising; reading the journals of the time, one could be excused for thinking the sole purpose of the colony was to maroon a military garrison in hell. There wasn't much ink wasted on the convicts either.

The journals and the letters home share a common theme of unimpressed amazement at the sights of the strange new land. Everything about New South Wales was entirely different from anywhere the British had been. It was as if they arrived on another, almost habitable, planet.

When a sheet of canvas is all you have for a home there are more important things to write about than the daily hijinks of the local dogs. For the modern historian, more's the pity, but a clear picture of the colony's dog life can still be formed. That's because environment, circumstance and instinct always influence canine behaviour, and because dogs will be dogs, they react predictably to situations and stimuli.

The canine novelty of Sydney Cove was the dingo. That novelty would soon wear off.

Dingoes were quite common around the site of the first settlement. Several journalists of the day noted their presence soon after the First Fleet arrived. They were usually seen in

the company of the Eora, who at first were thought to own all of them, the wild dingoes included.

Governor Arthur Phillip's journal had several contributors, so it is unclear who wrote what, but the dingo gets a mention in a paragraph titled 'Dog of New South Wales'.[8] The author provides a reasonably accurate description and considers the dingo to be fox-like, but larger. However, the accompanying drawing looks like a cross between a fox and a dingo; its sway-backed body is too long, its tail is as thick at the butt as a kangaroo's, and the diagram of the jaw looks like a representation of the lower mandible of a prehistoric reptile, with two sharply curved, pointy canines.

Watkin Tench was a marine lieutenant with the First Fleet and left England with a contract to chronicle the journey to New South Wales and the establishment of the new colony. His *Narrative of the Expedition to Botany Bay* is an all-round view of colonial life and the landscape of New South Wales, and the dingo is the first native animal he mentions. The dingo impressed Tench with its kangaroo-hunting skills, and he was equally struck by the tactics the kangaroos used to escape capture. No chase was a foregone conclusion either way.[9]

John White was the colony's surgeon general. He published an account of the new colony entitled *Journal of a Voyage to New South Wales*.[10] Elsewhere he described New South Wales as 'a country and place so forbidding and so hateful as only to merit execration and curses'.[11] Charged with dealing with the sick, injured and dying under the most primitive and trying circumstances, he obviously disliked New South Wales, and it's hardly surprising his journal is somewhat cheerless.

'The natives had with them some middling-sized dogs,' he wrote on 21 July 1788, 'somewhat resembling the species

Dog of New South Wales

A depiction of the dingo from Sydney colony surgeon John White's *Journal of a Voyage to New South Wales.* (Courtesy of National Library of Australia)

called in England fox-dog.' In eighteenth-century England, the keeshond and the Pomeranian were referred to as 'fox-dogs', because of their superficial resemblance to the fox.

> A servant of Captain Shea being one day out shooting, he found a very young puppy, belonging to the natives, eating part of a dead Kangaroo. He brought it to the camp, and it thrives much. The dog, in shape, is rather short and well made, has very fine hair of the nature of fur, and a sagacious look. When found, though not more than a month old, he showed some symptoms of ferocity. It was a considerable time before he could be induced to eat any flesh that was boiled, but he would gorge it raw with great avidity.[12]

Seeing these animals in the company of the Eora, John White assumed all dingoes were domesticated. Had there been fluent communication between the two peoples, White would have learned that the Eora, like all the other Aboriginal groups, only ever borrowed them. The behaviour of the little puppy and its unfamiliarity with cooked meat clearly indicate it was a wild dingo unused to human contact. That it was feeding on a carcass also reveals that the kangaroo had been killed by the puppy's own pack, not the Eora. They would never kill a kangaroo and just leave it to rot.

George Worgan served as a surgeon on the *Sirius*. He recorded the journey to New South Wales and the founding of the colony in his journal. He notes one of the earliest sightings of the dingo, when a group of about twelve Aboriginal people were encountered on the shore near where the *Supply* was anchored. They had a dog with them, 'something of the fox species'. Worgan observes that the Eora's dingoes were 'very tame and domestic'.[13]

Worgan's captain, the future second governor, John Hunter, had his own take on the dingo. He captured several as puppies and raised them as domestic dogs, but he never could cure them of their natural instincts:

> I had one which was a little puppy when caught, but,
> notwithstanding I took much pains to correct and cure it
> of its savageness, I found it took every opportunity, which
> it met with, to snap off the head of a fowl, or worry a pig,
> and would do it in defiance of correction. They are a very
> good-natured animal when domesticated, but I believe it
> to be impossible to cure that savageness, which all I have
> seen seem to possess.[14]

Watkin Tench noted that the dingo was the Eora's only domestic animal. He too thought it looked like the 'fox dog' of England, and stated that the Eora called it 'dingo'. (The appellation 'dingo' seems to have come into currency quite soon after the settlers' arrival.) The dingoes were attached to the Eora, Tench said, but disliked the colonists, though he also records that Governor Phillip took possession of a dingo puppy that apparently bonded with its new owner.

Tench tells a story, surely apocryphal, that the 'Indians' – as Aboriginal people were often called in the colony's early years – encouraged their dingoes to attack unaccompanied colonists they met in the bush:

> A surly fellow was one day out shooting, when the natives
> attempted to divert themselves in this manner at his expense.
> The man bore the teasing and gnawing of the dog at his
> heels for some time, but apprehending at length, that his
> patience might embolden them to use still further liberties,
> he turned around and shot the poor dingo dead on the spot:
> the owners of him set off with the utmost expedition.[15]

As you would – but a dingo puppy attacking on command? Not likely, mate! Someone had practised upon Lieutenant Tench, but that was nothing unusual; there were a few tall stories getting around. Convicts had reported seeing tigers and other monsters in the bush. The same stories of mythical creatures at large would still be in currency 200 years later, as we'll discover.

The convicts thought China, and freedom, lay little more than 100 miles away to the north across the Hawkesbury River. Some Irish convicts in particular loved that tall tale.

A convict named John Wilson absconded and toughed it out in the bush for a while, living for a time with the Indigenous inhabitants. Not making it all the way to China, if he was headed that way, he decided flogging was a better outcome than a life alone in the scrub. He came back and provided the authorities with a description of a creature he had seen, saying it was 'larger than a dog and, its hind parts thin, and bearing no proportion to the shoulders, which were strong and large'.[16]

As many an adventurous young Sydneysider can tell you, after rain the lush river flats along the Hawkesbury produce a bounteous crop of hallucinogenic, gold-topped mushrooms. Perhaps Mr Wilson turned to them for sustenance.

Returning to more credulous ground, John White recorded the first instance of emu hunting in colonial New South Wales,[17] and noted that greyhounds were unable to run down the first emu (or 'New Holland Cassowary') they pursued. Greyhounds are sprinters and need to capture their quarry relatively quickly. That lucky emu must have had a decent head start to be able to outpace them. It was the first mention in the colony of hunting with greyhounds, dogs that would make a great contribution to the colonists' survival.

John White was Australia's first serious naturalist. His journal contains a series of reasonable colour renderings of native plants, animals and birds. The dingo gets its own colour plate. A description titled 'A Dingo, Or Dog, of New South Wales'[18] follows. Its body looks fairly reasonable, other than its European red-fox colouring, and the fact that its tail is too low down. But its head! From the neck up, the poor fellow looks like a misshapen, snarling fox with cropped ears.

The illustrated dingo's nasty countenance suits Mr White's unflattering assessment. He calls it 'a variety of the dog' and likens it in size to a shepherd's dog, but says in other ways it resembles the wolf. The ears are short and erect, he writes, and the tail bushy. White had obviously only seen tan-coloured dingoes to that point. He describes the colour as a reddish dun colour, and the coat as long and thick, but straight.

Already the dingo had developed a nasty reputation:

> It is very ill-natured and vicious, and snarls, howls, and moans, like dogs in common. Whether this is the only dog in New South Wales, and whether they have it in a wild state, is not mentioned; but I should be inclined to believe they had no other; in which case, it will constitute the wolf of that country; and that which is domesticated is only the wild dog tamed, without having yet produced a variety, as in some parts of America.[19]

The colony's dissatisfied chaplain, the Reverend Richard Johnson, took a breather from whingeing to the governor about the licentiousness of the convicts and not having a church, to write to Evan Nepean the Permanent Under-Secretary of State for the Home Department in England. After mentioning the kangaroo and lesser marsupials, he too described the dingo as being fox-like.[20]

David Collins, the colony's judge-advocate, had a less flattering and typically insensitive take on the dingo. He believed them to be a species of jackal, and considered that 'they have an invincible predilection for poultry, which the severest beatings could never repress'.[21] Domestic fowls in a squawking flap are predator magnets, capable of flicking the attack switch in even the most docile domestic dog. Beating up a dingo for what comes naturally would suppress neither its predatory drive, nor its hunger. Nor would keeping it in miserable captivity turn it into a domestic dog. Collins did, however, begrudgingly admit they were a handsome breed.

It is very surprising that the dingoes' nocturnal howling was not more widely recorded; the mournful dirge would have been the predominant sound of the night, and would have given the wolf-allergic British the creeps. The domestic dogs would have joined in, particularly the hound types that are howling-prone, so perhaps no one took much notice of it.

It is apparent that dingoes were common in the vicinity of the new colony – mainly the tan dingoes, but also the black and tan, and the rarer black dingoes. The way things were going, the colonists would have a lot more dealings with the creatures before too long.

* * *

An assortment of New South Wales's faunal oddities would sail off with Arthur Phillip when he took his leave of the colony in late 1792: as David Collins notes in his journal, four kangaroos, several dingoes and a collection of other animals joined the journey.[22] Phillip also took his Aboriginal mate Bennelong with him.

The dingoes were presented to prominent personages in England. Phillip's reminiscences paint an ugly picture of a specimen given to Evan Nepean. He states that it was particularly savage and would attack any animal it saw. It attacked a donkey and other dogs, and was fond of killing poultry and rabbits.[23] Regrettably, this one unhappy, disenfranchised, mismanaged individual was a poor ambassador for the dingo.

Two years later, in 1800, the illustrator Sydenham Edwards produced *Cynographia Britannica*,[24] an encyclopaedia of British dog breeds. The dingo gets his own little chapter as 'Canis Antarcticus – The Dingo, or Dog of New South Wales'. Mr Edwards provided an accurate description and admitted that dingoes in captivity were 'pretty docile, but have not shown any attachment to their keepers'. He had learned from his friend, Major George Johnston of the New South Wales Corps, that dingoes whelped their litters in fallen hollow trees, and could be tamed, but not domesticated.

That wasn't really such a bad rap after the previous poor publicity.

CHAPTER 3

Colonial Hounds Save the Day

Within months of the colonists' arrival they faced the prospect of starvation. Everyone and everything needed feeding. Survival was going to be a near-run thing.

As Tench noted, most colonists would do anything to secure food. Tench and his fellow officers, who were no better fed than the convicts and ordinary soldiers, were not above stealing a dingo's kangaroo kill just to survive.[1]

The colonists' first attempts at growing crops failed, and soon, because of a lack of fruit and vegetables, the effects of scurvy appeared among all classes of the people. Better farming land became the highest priority, but it lay well away from the thin, sandy infertile soils of the harbour.

In a new settlement without a stick of infrastructure, keeping the weather off the food and equipment was also a priority. Initially the colonists used tents for storage, then they began to construct huts thatched with bulrushes and the fronds of the cabbage-tree palm.

There were thieves at every turn. Fatigued, hungry marines, suffering from exposure and every other attendant torment of temptation, were left in charge of stores security, which was akin to leaving Count Dracula in charge of the blood bank. Six marines faced a court martial in 1789 and were subsequently hanged for stealing from the government stores.[2] Governor Phillip had the same issues with theft from the colony's garden.

David Collins complained that a starving party pooper had nicked a sheep being fattened for an officers-only party. Even though the governor offered freedom for anyone prepared to inform on the thief, the crime went unsolved.[3] Considering the small size of the colony at the time, it was quite a successful heist.

Protection of the stores and crops was an imperative that could have been better effected with the aid of dogs. The English had produced some effective crime-deterring breeds, such as those used on His Lordship's estate – the English mastiff and bull mastiff – as well as the ban dog. Perhaps posting mastiffs as sentries around the storehouses and gardens would have proven an effective impediment. As we'll see, dogs were certainly used successfully to hinder convict escapes in Van Diemen's Land (Tasmania) not long afterwards.

But it wasn't just the hungry convicts and marines who attacked the stores. Tench recorded that rats soon became a serious threat to food stores and the feeble vegetable plots.[4]

The old world had maintained a hate–hate relationship with the rat for centuries. After all, it carried the nasty little flea that spread the bubonic plague. But far from being left behind in England, the rat had hitched a ride with the First Fleet in sufficient numbers to cause major problems in New South Wales.

Elsewhere in his journal, David Collins mentions that some officers brought terriers with them.[5] There were also cats on the First Fleet. It appears these animals were put to work rat killing, or did so themselves through instinct or hunger.

Rat kangaroos and other small native herbivores found themselves quite partial to young, sweet vegetables, and would also have contributed to the assault on the first gardens. The native rat kangaroo and imported rat populations exploded. More terriers were brought out from England later, but by then the European rats had gained a foothold in Sydney that they would never relinquish.

* * *

The colonists also couldn't adequately control or protect their limited livestock, which, just like their spindly crops, were constantly ravaged by theft and predation.

Things had got off to a bad start for Major Robert Ross's own private stock, when lightning struck six ewes and two lambs and an unlucky pig on the night the First Fleet's female convicts shuffled ashore.[6] Surely that pre-cooked meat didn't go to waste, but this misfortune was just the beginning.

The First Fleet had embarked Cape fat-tailed sheep at the Cape of Good Hope and brought them to one big, open, dangerous paddock. Phillip had no experienced or reliable shepherds or stockmen on the books, and when pressed into service convict shepherds were not of much use to anyone. Despite being guarded, the flock was soon decimated.

The first record of predation on domestic stock is a journal entry by the governor,[7] after he had undertaken an exploration of Broken Bay and the Hawkesbury River by boat, in search of better agricultural land. The trip was a success. Extensive fertile river flats were found that would ultimately help support the colony, but the governor returned to find that five ewes and a lamb had been mauled and killed in broad daylight.[8] You wouldn't envy the convict shepherd who let that happen on his watch.

For a nation of dog-breed creators, the English were very short on sheepdogs when they arrived in New South Wales. It also seems strange Arthur Phillip did not anticipate that guarding dogs would be needed in New South Wales, where he knew there were wolves. Sheep-guarding dogs would have provided excellent protection from dingoes and domestic dogs, as they do throughout Australia today. As it was, sheep numbers were dwindling, and predation by dingoes was the last thing Governor Phillip needed.

For millennia, dogs had been man's vital tool in controlling and protecting domestic stock. Yet it appears the first dogs came to New South Wales more as companions, or as a sporting afterthought.

It is astounding that Arthur Phillip, an experienced farmer, should seek to establish a colony supported by agriculture in a wild, foreign land, but should not bring working and stock-guarding dogs with him (or draught animals, for that matter).

* * *

Along with the sheep, Phillip had embarked seven black Cape cattle at the Cape of Good Hope, two bulls and five cows, one of which was later shot for being recalcitrant and difficult, or dangerous to manage. They were a good choice for the colony, being hardy and from a reasonably similar climate. Cape cattle were the progenitors of the impressive, medium-sized Drakensberger, a South African breed that, like its forebear, is now making its way into Australia.

The small mob of six cattle were grazed in the area now known as the Domain and yarded at night at the 'farm' on the site of today's Royal Botanic Garden. They were put under the care of a convict named Edward Corbett, who was instructed not to lose them. Governor Phillip had warned all the convicts that 'the life of a breeding animal was worth a man's'.[9]

How one man was to manage six head of cattle alone, without the aid of suitable dogs, is anyone's guess. In early June 1788, Edward Corbett left the cattle alone briefly and nicked off for his lunch. When he returned, the two bulls and four cows had taken their leave of the colony. Edward Corbett, remembering the draconian terms and conditions of his employment, promptly did likewise.[10]

Phillip sent search parties out looking for the cattle and Corbett, who was thought to have stolen them. After days of fruitless searching the governor wrote the cattle off as having been speared by the Eora, but continued the hunt for Corbett. To cut a long story short, Edward Corbett was later found starved, emaciated and near death. He was hospitalised, restored to spry good health, and punished for losing the cattle. He was eventually hanged, this time for knocking off a smock.

Nothing more was heard of the cattle until they were discovered by two convict kangaroo hunters in 1795, at a place subsequently called the Cowpastures, very near what is now Camden. They numbered around 100 by this time and, left to go forth and multiply, their numbers had risen to about 4000 by Governor Macquarie's era, around 1815.

Six runaway sheep would not have lasted a week in the bush. It was their size and aggression that had allowed the cattle to prosper. Their increase also demonstrated they could coexist with dingoes.

When the cattle were found in 1795, Governor John Hunter and two other officers rode out to the Cowpastures to inspect them. The now-wild cattle objected to the intrusion,

and an enraged bull and his harem tried to attack Governor
Hunter and his aides, one of whom was the naval surgeon and
maritime explorer George Bass. The cattle were only
persuaded to 'desist and decamp' when dogs (presumably
greyhounds) were set on them.[11] That aggression may have also
convinced the Aboriginal people looking to prey on them that
there was easier and less lethal meat to be had.

The upshot of the entire wild-cattle saga was that Edward
Corbett's rumbling stomach had indirectly shown that beef
production had a bright future in New South Wales. In
retrospect, he probably should have been granted a ticket-of-
leave for his services to the colony. Hindsight is an exact
science.

* * *

With limited and deteriorating food supplies, failing crops and
dwindling livestock, the colonists found themselves in the
unlikely position of having to live off the land to supplement
their rations.

Fish were the obvious fall-back option, but the convicts
preferred the disgustingly rancid salt beef or pork to the fresh
fish of the harbour, and by the onset of winter all the coastal
waters proved to be very quiet fishing-wise. The shortages had
got so bad that they turned to scavenging and eating native
plants to ward off scurvy.[12]

Such instances demonstrate how inept the British were in
not only provisioning the expedition, but understanding and
adjusting to their surroundings.

The Eora followed the food as season and availability
dictated. They were doing all right – other than the fact that

they were dying in droves from smallpox brought by the British.

The colonists concentrated around Sydney Cove, on the other hand, were comparatively great in number, and everything about their sedentary lifestyle disturbed the natural order.

When they arrived they hadn't even bothered knocking. They'd just marched into the Eora's home, oblivious to any consequences, and set up shop.

Turning up uninvited and unannounced was one thing, but not bringing enough food for man or beast was quite another. With typical British arrogance, they'd thought they would just set up bounteous farms and establish a new England. Yet the New South Wales they found was an unpredictable and tight-fisted mistress, who refused to help these appallingly under-prepared newcomers.

It was a big, lanky, mongrel dog that provided the First Fleeters with a wholesome form of fresh protein until the colony could be resupplied.

* * *

The local kangaroo, the eastern grey, was an obvious alternative source of protein. Kangaroo meat is very healthy, but gamey, and quite dry compared with the fattier meats that Europeans are accustomed to, like lamb or beef. Even the Eora preferred fish. There are no fat kangaroos, and they don't carry a lot of meat above their massive hind legs. The rest is mostly just skin and bone, and suitable for soups and stews. But when you're hungry enough, anything will do, and none of the famished colonists complained about it.

Phillip sent marines and the colony's best marksmen and most trusted convicts out to hunt kangaroos under the strict supervision of a 'trusty sergeant'.[13]

The grey kangaroo is a creature of the woodlands, but feeds out in relatively open country. Stalking a kangaroo effectively is a lengthy, painstaking process, because those worth hunting are the small does and immature bucks, and they always mob up when grazing. Lookout duty is shared, and one is always upright keeping an eye out. After many millennia of predation by Aboriginal people, kangaroos had developed an understandable distrust of people.

Hunting kangaroos with muskets was usually unsuccessful. The Brown Bess muskets used by the British had a maximum range of just over 150 yards, but were reliably accurate at more like 50 yards, maybe less. The hunters tried hiding in cover on the edges of the open country and ambushing their quarry, but after several futile forays they concluded – correctly – that hounds were their best chance.[14]

Coursing kangaroos is fraught with danger. Kangaroos look Skippy-cute, but they are not to be messed with, especially the large bucks. The dingoes had long learned to leave the big, healthy fellows alone. A hunter with a weapon can select his target; a dog is far less discriminating, and the English greyhounds had a lot to learn.

When pursued, eastern greys do not seek safety in numbers, but confound their pursuers by scattering in all directions. They make straight for the roughest country to hand, or for water. A kangaroo bounds effortlessly over all manner of dangerous obstacles that a dog must run through at breakneck speed, which is extremely dangerous for any dog. At Botany Bay back in 1770, Joseph Banks's greyhound was lured into

rough country, outsmarted by a joey-sized macropod, and promptly ran into a concealed stump, laming itself.[15] It was lucky it didn't do itself even greater damage.

When a kangaroo bails up it means business. With its tail taking its weight, it strikes with its powerful legs; the huge middle toes are a tool that will disembowel any dog careless enough to get within striking distance. Even the smaller does can be lethal.

When kangaroos reach water, they stand in it up to their waist and wait for a dog to come to them. Then they either use their feet to tear the dog open, or stand on top of the dog and drown it.

Dingoes, having thousands of years' experience, were highly skilled at killing kangaroos. It would take practice for the colony's greyhounds to master effective kangaroo-catching techniques. If the chase occurred in open country the hounds would usually prevail, but even when successful, they were often torn up and killed.

The colony's second governor, Captain John Hunter, was a keen observer of wildlife, and was impressed by the strength of the kangaroos when assailed by greyhounds.

For some time, he wrote, the colonists considered the kangaroo's tail as its chief defence. But Hunter hunted them himself with greyhounds and found out, at his dogs' expense, that kangaroos ripped with their hind legs, scratched with their 'hands' and bit savagely:

> As soon as the hound seizes him, he turns, and catching
> hold with the nails of his fore-paws, he springs upon, and
> strikes at the dog with the claws of his hind feet, which are
> wonderfully strong, and tears him to such a degree, that it

has frequently happened that we have been under the necessity of carrying the dog home, from the severity of his wounds: few of these kangaroos have ever effected their escape, after being seized by the dog, for they have generally caught them by the throat, and there held them until they were assisted, although many of them have very near lost their lives in the struggle.

The size of some of the big bucks astounded Hunter. Some of the male kangaroos he had seen were six feet tall when sitting on their haunches. A kangaroo that tall extends to well over seven angry feet when on the defensive, and on its toes:

Such an animal is too strong for a single dog, and although he might be much wounded, would, without the dog had assistance at hand, certainly kill him. We know that the native dogs of this country hunt and kill the kangaroo; they may be more fierce, but they do not appear to be so strong as our large greyhound.[16]

Hunter believed the greyhound was much faster across the ground than the kangaroo, providing the chase occurred in open country. If the dogs got a clear run, he wrote, the chase rarely lasted more than ten minutes.

Surgeon John White – usually a reliable chronicler, but a sucker for a tall story – recorded the words of an anonymous fibber who told him with a straight face that some convicts went out kangaroo hunting with Hector the Newfoundland dog. Displaying athletic prowess unknown to Newfoundlands, Hector allegedly fastened onto a large kangaroo. Unable to shake off its assailant, the kangaroo beat up Hector using its

tail, causing him to be 'severely bruised, and cut in several places'.[17]

Hector the Newfoundland catching kangaroos? Oh, come on! Newfies are the world's gentlest dogs – big, loveable, lumbering things. They originated in eastern Canada as a general-purpose fisherman's dog; with their webbed feet and love of the water, they were used to carry warps and ropes from vessel to vessel. They have always had a reputation for saving drowning people and were certainly being introduced to Britain by the time of the First Fleet.

Newfies are known for their love of people, especially children. Hector was obviously typical of his breed, being famously devoted to his master, Mr Marshall, who apparently treated Hector a lot better than he did his fellow man. Marshall was the infamous master of the *Scarborough*, which, along with the other three convict transports of the Second Fleet, delivered what remained of its surviving human cargo to New South Wales in the most wretched state.[18]

Thousands of people in Britain with an earnest interest in New South Wales would have read and believed John White's ridiculous story, which even Governor Phillip repeated in his journal. Things are crook when even the governor gets sucked in. As we'll see, this was just the beginning of a robust tradition of stubborn myths regarding Australia's dogs and native animals.

The power and ferocity of docile kangaroos pushed to self-defence impressed Watkin Tench too. He noted that they were fast across the ground and capable of bounds exceeding 20 feet. Like Hunter, he believed the greyhounds were faster, but had great difficulty in subduing kangaroos singly or if the protection of scrub lay close to hand:

> The greyhounds for a long time were incapable of taking
> them; but with a brace of dogs, if not near cover a
> kangaroo almost always falls, since the greyhounds have
> acquired by practice the proper method of fastening upon
> them. Nevertheless, they often miserably tear the dogs.
> The rough wiry greyhound suffers least in the conflict, and
> is most prized by the hunters.[19]

The smooth-coated greyhound was nearly, but not quite, the dog needed to consistently and effectively tackle the kangaroo. While fast and game enough, it was not built for tackling large prey, or coursing in scrub. The greyhound is a finely built, open-country sprinter, with a short silky coat, commensurately thin-skinned, and thus too easily damaged. It is the nature of eucalypts to drop limbs. Hidden stumps can break a dog's neck, and dogs colliding with broken branches on standing and fallen trees were often mortally staked.

The passage above is the earliest account of the colony in which the wire-coated greyhound is mentioned. Wire-coated dogs are thicker skinned than their smooth-coated relatives, and have been developed for hunting dangerous game in thick conditions. Thicker skin also means tougher feet. The hot, dry ground of Australia, littered with all manner of sharp hazards, was hard on the feet of light-soled dogs.

Joseph Banks, by now an influential patron of the natural sciences, would surely have spoken to Phillip and advised him to take dogs suited to the tough conditions. The 'rough wiry greyhound' Tench mentions and the 'large greyhounds' cited by John Hunter may have been the now-extinct Scotch rough-haired greyhound, a lighter version of the Scottish deerhound, but heavier than the greyhound.

Hugh Dalziel describes the Scotch rough-haired greyhound in his 1879 book *British Dogs*: 'A Deerhound–greyhound cross, the shape of the rough greyhound corresponds closely with the Deerhound; but he is not so large and powerful, averaging about twenty-six inches at the shoulder against thirty inches in the Deerhound.'[20] It was basically the British version of the kangaroo dog.

Dalziel says the Scotch rough-haired greyhound is also known as the wiry-haired greyhound and is larger-boned than the smooth greyhound, not quite so elegant in shape, and 'wanting in that beautiful finish that stamps the modern greyhound'. In other words, ugly. 'The rough, harsh coat adds to the effect, and the hairy jaws make the head being wider between the ears, which are also apt to be large and carried in an ugly manner.'[21]

It is certain that the Scotch rough-haired greyhound or the Scottish deerhound, possibly both, arrived in New South Wales with the First Fleet. The Scottish deerhound shares a common ancestry with the Irish wolfhound. It was originally a larger type of Scottish wolfhound, but when the wolf was driven to extinction in the United Kingdom, its job description changed. It gradually decreased a little in size so it could run faster and further, and became the Scottish deerhound. It also seems to have been present in either form – deerhound or rough-haired greyhound – in New South Wales from the earliest days of colonisation, but references to which breed are hazy and confusing. The problem is that the journalists of the day may not have used the correct appellations for the dogs they were referring to.

The powerful, heavily built and thick-skinned Scottish deerhound found favour with the colony's hunters, though for kangaroo work it would need to be quicker off the mark. So the colonists set about creating a dog better suited to local conditions.

They crossbred Scottish deerhounds or possibly Scotch rough-haired greyhounds with smooth-coated greyhounds, and in doing so produced mixed litters of wiry- and smooth-coated puppies, the wiry-coated being much preferred. The offspring were closer to the deerhound in size – 80 pounds (36 kilograms) as opposed to the smooth-coated classic greyhound, which was 65 pounds (29 kilograms). They retained both the greyhound's explosive start and speed, and the deerhound's greater stamina for long chases.

These offspring came to be known as kangaroo dogs, but it appears that in the early days any running sight hound in New South Wales was called a 'greyhound'. The kangaroo dogs were perfect long-distance runners, thick-skinned with damage-resistant coats, and they were powerful. It was these dogs that would supplement the colony's rations and ward off the threat of starvation.

Kangaroo dogs kept the deerhound's large-game-hunting genes, and they caught on quick. Working in pairs, they devised an effective method of catching and killing a kangaroo. For it to work, they needed to reach their prey before the kangaroo had time to reach rough country or bail up.

First one dog, the faster of the two, usually a bitch – or as they were called in those days, a slut – would position itself close beside and slightly behind the kangaroo. The dogs had their favoured side to work from and were either right- or left-handed. As the kangaroo was on the upward leap – at speeds of up to forty kilometres per hour, mind you – the lead dog would seize it by the butt of the tail and drive it headlong to

the ground. The following dog, usually the larger, stronger 'dog' – the term for a male – would then leap onto the kangaroo from behind and break its neck.

It was the safest method, far safer than confronting a kangaroo that had bailed up and could grasp the dogs and then disembowel them with their formidable toes, but still extremely dangerous for the hounds. Once a kangaroo is deprived of the full and natural use of its tail it is vulnerable, but the dog holding the tail could not afford to lose its grip or get in the way of the kangaroo's lethal hind legs. Kangaroo dogs were often maimed or killed by kangaroos, and every working hound carried scars, or 'honourable medals', as they became known.

As with any other creature, not all kangaroo dogs were equal. Some were bigger, some were faster, some were more aggressive, and some had better technique. There is no record

Kangaroo dogs were quickly appreciated by Aboriginal people living traditional lifestyles. They made hunting kangaroos much easier. This man from Australia's interior is pictured with three kangaroo dogs. (Courtesy of State Library of Western Australia)

of the selective breeding of kangaroo dogs in early New South Wales, but there is no question that it would have happened. It does with any domestic animal.

Successive British ships brought more smooth-coated greyhounds and deerhounds, and within a few years, the quality and ability of the kangaroo dogs would have increased. Puppies would have been born with an instinctive ability to course and kill kangaroos, a specialised and dangerous prey creature if there ever was one.

As we've learned, the British were weaned on blood sports. The highest-profile non-human sports figures in England at the time were champion fighting or bull- and bear-baiting dogs. It would have been no different in New South Wales. The reputation of champion kangaroo dogs would have meant prestige and big money (or a lot of rum, in the rum economy) for their owners. The kangaroo dog became the most valued dog in the colony.

Kangaroo-dog breeding became big business in New South Wales and supply could not meet demand. As new colonies were established in Van Diemen's Land and throughout the mainland, the kangaroo dog became necessary for pastoral success. Its reputation spread far and wide, and many were exported in later years to India and South Africa for antelope coursing. Their ability to work in trying conditions made them extremely popular.

Like all their sight-hound kin, kangaroo dogs also had a kindly disposition and a noble bearing. Sight hounds have always been known as serene companions. For such a large breed, the kangaroo dog curled up into a surprisingly compact ball, did not have a massive appetite and, like all the sight hounds, slept when it wasn't on duty.

The kangaroo dog has generally been considered a breed, as for over 100 years these dogs were generally similar in build, but in truth they were more a type. Ability, not appearance, is the prerequisite in any dog when its work is of such crucial importance. It wasn't until the dog show became popular in the mid- to late nineteenth century that much emphasis was placed on a dog's pedigree history or its appearance.

* * *

Thanks to the kangaroo dog, the colony's food problems were alleviated. But there is a simple rule to hunting pressure: the more you hunt, the more scattered and wary the game becomes, and the further you need to go to hunt. As the colonists turned their attention to kangaroos, they were forced to go further and further afield to find them. That created a major problem for the local dingoes, who so far hadn't proven to be too troublesome. But now, the packs surrounding the colony were left without their natural large prey.

Something had to give. The dingoes turned their attention to the other obvious food source within their territories: domestic stock. And where the dingoes went, feral and domestic dogs followed in a typically canine spirit of collaboration that meant nothing but trouble.

CHAPTER 4

Sydney's Dogs Behave Badly

The colonists never realised it, but it was they who fired the opening salvo in the dingo wars that rage to this day. They plonked themselves and their vulnerable livestock right into the territory of a dingo pack. Then they eliminated the dingo's natural large prey. They invited hordes of rat kangaroos and their own imported rats to their stores and gardens, and where there is prey, there will be predators. Their sheep didn't have to go anywhere to be exposed to the dingoes – the dingoes came to them, because now there was plenty of opportunity for an easy feed around the growing, unprotected camp.

In the dingo, the colonists were dealing with a predator with a bold tolerance for human proximity. Dingoes were used to exploiting people. They weren't shy like an old-world wolf, because they hadn't yet learned the wolf's fear of man. But the dingoes would soon discover that not all men were as tolerant and understanding as the Eora.

* * *

Rose Hill up the sharp end of the Parramatta River – today's Parramatta – was becoming a successful choice for farm sites. It was there that ticket-of-leave man James Ruse grew the colony's first successful private wheat and corn crops in 1790, and that John Macarthur and his indomitable wife, Elizabeth, first began breeding merino sheep in 1797.

Initially Rose Hill was only easily accessible by boat, the country between Sydney Town and Rose Hill being difficult to negotiate, until a rough track, the Parramatta Road, was built through the scrub. It's hard to imagine that the stop-start, fume-ridden Parramatta Road of today, lined with car yards and super stores, was once so primitive. Today there is not one eucalypt along its frantic, fumy length, but back in the day the bush all around Sydney Town and Parramatta was dotted with the isolated bark huts and subsistence farms of free settlers.

Many were so poor they could only grow a few vegetables, and maybe they kept poultry. Others perhaps kept sheep and pigs, but their stock was preserved for breeding. Either way, a major source of protein was the kangaroo, and since few settlers had firearms, their main means of obtaining it was the kangaroo dog.

The big grizzled hunters became an indispensable tool for every settler. As well as helping keep farmers fed during these early, hungry years, they offered some measure of protection against marauding dingoes and attacks by the disenfranchised prior occupants.

Kangaroos were also becoming a menace for some settlers. Back then kangaroo numbers were not as large as they are today, now that protection, improved pasture and reliable water are available to support them. It is the nature of any animal to make the easiest living it can. Young, tender food crops were an irresistible lure for an animal doing it tough. Kangaroos were nocturnal feeders, and most of the damage to crops occurred when people had no chance of protecting their livelihood. As land clearing increased and farmland proliferated, the kangaroo, benefiting from the abundance of food, developed pretty expensive tastes. It became a serious destroyer of crops, and a competitor with sheep and cattle for grasses.

It took a couple of years for even the best-managed farms to become self-sufficient. Elizabeth Macarthur managed the family's extensive holdings – she was the brains of that outfit – while her husband, John, was banished to England from 1801 to 1805, and again from 1809 to 1817. She mentions in a letter to a friend the worth of owning greyhounds in the early days of their first farm at Rose Hill. She did not elaborate on whether her greyhounds were the classic smooth-coated or the kangaroo-dog types, but wrote that her stock consisted of one horse, two mares, two cows, 130 goats, upwards of 100 pigs, and an abundance of poultry. She had received no stock from the government beyond one of the cows; the rest she had either purchased or bred. 'With the assistance of one man and half a dozen greyhounds, which I keep, my table is constantly supplied with wild ducks or kangaroos. Averaging one week with another these dogs do not kill less than 300 lb. weight.'[1]

Three hundred pounds of kangaroo meat a week is an enormous number of animals killed in a local area. The kangaroo dogs were more than efficient in the open country that ran to the foothills of the Blue Mountains. Every settler would have been doing the same, and for several years, government hunters pursued kangaroos to keep the burgeoning convict and military population fed.

With such wholesale slaughter of kangaroos, it's little wonder dingoes started predating on the settlers' stock. They were beginning to cause serious problems of a night, when they came in large packs and preyed on unprotected and even yarded stock. The colonists were fighting a war on three fronts now, because the Aboriginal people were also developing a liking for mutton and lamb, not to mention crops such as maize (sweet corn).

In July 1804, a parliamentary committee at the Council Chamber at Whitehall in London questioned John Macarthur about the prospects of the wool industry in New South Wales. They asked: are there any animals in New South Wales destructive to sheep?

Macarthur replied: 'None except the native dog, which is an animal somewhat between a fox and a wolf. There are not many of them, and they are so timid in their nature that they will not approach the sheep by day.'

Macarthur deliberately understated the dingo problem to fast-track support and approval of his wool-growing ventures. His daughter Elizabeth recorded that '[Her father's] flocks suffered much from the depredations of the native dog.'[2]

A year later, in Sydney, Rowland Hassall, a prominent sheep farmer and manager of the government farm, responded by letter to inquiries from London regarding the state of wool-growing in New South Wales. The question was: how long do you suppose it will be before your flocks will be increased to twice their present number?

Hassall was more pessimistic about the wool industry's chances of survival. He replied: 'This seems the most difficult to answer, as the wet seasons – the dishonesty and carelessness of the shepherds – the destruction that the native dogs often make as well as many other causes that might be mentioned argue much against their Increase.'[3]

Hassall was wrong on that count: the Macarthurs and the Reverend Samuel Marsden had already demonstrated that with well-bred merinos, New South Wales was ideal for wool-growing.

But Hassall was right about the predation by dogs. Typically, though, it was only the dingoes that got the bad rap, when much of the blame lay with plain old domestic dogs.

* * *

When the first predation of the colony's livestock occurred, Governor Phillip had blamed the local dingoes.

Since the attack occurred in broad daylight, the governor was probably wrong in this assumption. The local wild pack might have become bold enough, but you'd imagine it would attack at night. The broad-daylight assault had hungry domestic dog written all over it, and it was the first of the

The dingo's reputation for savagery reached ridiculous levels in Britain and Europe in the late eighteenth and early nineteenth centuries. This apocalyptic woodcut by Walter Hart features demon-faced dingoes running down a terrified horse and desperate horseman. (Courtesy of State Library of Victoria, IAN27/07/66/5)

frenzied mass killings of sheep that continue to plague the wool and fat-lamb industries in Australia.

The Eora's dingo puppies were the perfect scapegoats to finger for killing the governor's sheep. Considering the potential consequences, what owners of the dogs responsible, or what negligent sheep-guarding convict, *wouldn't* want to blame the Eora's dingoes?

Royal decree had ordered that the Eora were not to be mistreated, but welcomed into the warm, protective embrace of good King George and his empire. Yet other than Governor Phillip and a few of his more enlightened officers, everyone hated them. The convicts hated them because convicts had no one beneath them on the social ladder. The Eora were largely free to do as they pleased, and were not starving. The marines hated the convicts, naturally, and they hated the Eora for much the same reasons as the convicts did: because they too were prisoners to some degree.

The common response of owners of destructive dogs is to point the finger at anything or anyone but their own animal. (The prime example is Van Diemen's Land, where thylacines would be blamed for sheep kills committed by domestic and feral dogs.)

The difference in 1788 was that the dingoes had been pigeonholed as wolf-like and bad, and therefore were expected to kill sheep. Domestic dogs were good, and were not expected to kill like a wild dog, unless they were hunting.

In developing the dog, man had turned a nocturnal predator, the wolf, into a creature like himself that slept at night – or was supposed to sleep at night. Yet dogs without the constraints of human attachment and control can easily revert to nocturnal hunting.

If the colonists were hungry, so were their dogs. There is no doubt they needed to supplement their own meagre rations, just as their owners did. Any dog food would have been salt beef or pork rendered inedible by age, or leftovers from hunting excursions. There would not have been much to spare, and hungry dogs are alert, active dogs. An abundance of rats and rat kangaroos would have been a ready source of small, difficult-to-catch food. But any dog that needs to kill to survive won't discriminate. Sheep were a large, easy-to-find animal to kill.

Once a domestic dog experiences the uninhibited joy of mauling sheep, it might be physically prevented from doing it again, but it cannot be cured of the addiction. Inveterate nocturnal sheep killers have often been found to be meek, sweet-natured pets by day. It is simply the failure to keep dogs under effective control that leads to all such predatory behaviour.

Every town or city has a feral or nuisance dog problem. Sydney's began the day the First Fleet arrived, and would get a lot worse over the next fifty years.

Correctly managed dogs are without question an asset to any community. Dogs provide real companionship for people, and have always been the cheapest form of home insurance available. Even little dogs make enough noise to convince thieves to move on.

But dogs without effective control are a genuine liability. When a dog roams he will join other roaming dogs to form a pack. The dog pack is no different from any other group. A group provides certainty, and the safety and power in numbers greatly increases the chances of the individual's survival, but not necessarily their contentment.

Dogs in a pack conform to inflexible, hierarchical rules, under the guidance and protection of a leader. Dominant dogs aren't usually all that happy. Being the boss, for humans and dogs, is so encumbered by responsibility that it's not a lot of fun. Dogs lower in the order are usually more satisfied with their lot, until a weakness in the leadership becomes apparent, and that's when the troubles start.

The domestic dog pack is different from the wild pack, of course, because there is a human in charge, or there should be. A smallish, self-managed, domestic dog pack takes all the hard work out of owning more than one dog, providing there are no overtly aggressive types within the group and sufficient training and socialisation has been established.

Here are the basic domestic dog-pack rules. Allow the natural order to establish itself without human interference. Yes, there may be a bit of noisy wrestling and gnashing of teeth, but once the pack order is established it must be preserved. That means feeding and bestowing favours upon the pack in order of seniority. And, of course, the furry pack leader must be obedient to the overall pack leader – the human. That's all there is to it, but it has always been a technique easier said than done.

Without a human, or an effective human, to lead them, dogs are always more dangerous when they're in company. Packs of semi-feral dogs are sometimes the cause of attacks on people, usually children or the elderly. Roaming dogs cause traffic accidents, scavenge, and prey on pets and stock.

Back in the early days of the colony it made no difference whether the uncontrolled dogs and dingoes were owned by the colonists or the local Aboriginal people. Like urban feral dogs today, the domestic ferals back then – known as cur dogs – were part-time wild dogs: happy to hang out with the family by day, then packing together with their mates to go killing of a night.

Sheep and goats were the most vulnerable stock, but pigs and calves were also attacked. In town, poultry and domestic rabbits were regularly killed too. Scavenging dogs would get into garbage, raid meat sheds, fight among themselves and become hazards to riders and horse-drawn carriages. Things always end up ugly for those types.

* * *

As more convict transports arrived and the colony grew, more free settlers came to New South Wales to take up land. Emancipated convicts took up new places in society, and both Sydney Town and Parramatta were expanding rapidly. The British presence in New South Wales radiated out from Sydney, Parramatta and later the Hawkesbury like separate drops of oil in a glass of water, broadening and expanding until they converged as a mass of farms that obliterated the bush and the old ways.

Year by year the colony inched towards self-sufficiency, but infrastructure to contain dogs remained nonexistent: a sure recipe for trouble. Stock were increasingly subject to the depredations of uncontrolled, feral dogs and dingoes, which even harassed travellers. They became so numerous that they ran in large, dangerous packs. Conflict was inevitable.

David Collins makes note of the feral dog problem in *An Account of the Colony of New South Wales*. He writes that 'Much mischief had been done by them among the hogs, sheep, goats, and fowls of individuals.'[4]

Twelve years after colonisation, the dog population, and the subsequent domestic feral problem – uncontrolled pet dogs causing havoc – had increased to such an extent that a public order became necessary. The order, issued by Governor Philip Gidley King in 1801, attempted to restrict the number of dogs kept by each person to no more than were necessary for the protection of their house and premises. The order also mentions a domestic goat problem. Little is known about this issue, but goats are far more difficult to control than sheep. They are skilled and innovative escape artists and being grazers and browsers, they cause great environmental damage when left to their own devices. While the goat problem is not recorded anywhere else in records of the time, they must have become a source of prey that contributed to the feral dog problem. The order proclaimed:

> Several individuals having complained of the great decrease
> of their sheep and lambs by the curs with which this
> colony abounds, and the great damage done to gardens by
> goats ranging without a herd, and as the breeding stock of
> sheep is of the greatest consequence to the welfare of this
> colony, no person is to suffer any cur dogs to follow them,
> or any cart, wheelbarrow, etc. The Governor having given
> permission to those who have flocks of sheep to order their
> herdsmen to kill any dogs that approach them, and the
> owners will forfeit treble the value of any stock killed by
> them. Persons who keep cur dogs that are in the habit of
> flying at horses are to destroy them, otherwise they will be
> indicted as a nuisance. It is recommended to those who
> have more dogs than one (except greyhounds or terriers) to
> kill them, as a tax will shortly be laid on all cur dogs.[5]

New South Wales was a brutal, hard place. The convicts and, as time went by, the Aboriginal people, were viewed worse than stock animals. One could hardly expect the dogs would receive better.[6]

A dog always has been, and always will be, what you make it. No matter what the era, many people have trouble understanding the right management techniques for the dogs in their care. Allowing dogs to breed without restraint and roam at large was just asking for trouble. It still is.

It is apparent that in 1801, there was a clear distinction in social value between the cur dogs and the 'greyhounds' and terriers. Both the latter were relied on for their meat-hunting and vermin-destruction contributions. The cur of the time was a mixture of the dingo and every type of domestic dog present in the colony: interbred, ill-bred, and for the most part, feral.

Six years later the urban ferals were still at it. Governor Bligh, known for being a 'hard horse', issued the following proclamation:

> Whereas a number of sheep and lambs, the property of
> Government and others, in and about Sydney, have been
> worried and killed by Dogs belonging to individuals in this
> Town and its Vicinity; It is hereby ordered, that every
> Dog, of every description be immediately destroyed,
> except those of Officers and respectable Housekeepers.
> Kangaroo dogs and House Dogs, which are kept for the
> defence of the Premises of such Officers or respectable
> House-keepers, are to be kept chained up; and the
> Constables are hereby authorised to take every means in
> their power to carry this Order into execution. Any
> persons from the date hereof, who are known to
> keep Dogs contrary to this Order, will on conviction
> before two or more Magistrates forfeit the sum of Ten

Pounds, to be levied on their respective goods and chattels, half to the Informer, and half to the Orphan Fund; and further to be answerable for all damages incurred by the violation of this Order.[7]

Whatever! Australian bureaucracy has always been out of touch with canine behaviour and how to manage it. Governors King's and Bligh's respective proclamations of 1801 and 1807 sounded tough, but in effect they did nothing to reduce the feral-dog problem. They were the first of many impractical and unenforceable dog-management strategies in Australia. The problem is that dog laws are usually framed by bureaucrats and enforced by public servants with no understanding of dog behaviour or efficient dog management.

News of Sydney's escalating dog problems even made it back to England. *Simmonds's Colonial Magazine and Foreign Miscellany* of May 1845 noted in its news snippet column 'Colonial Intelligence' that the dog problem in New South Wales had continued despite the government's draconian dog-control edicts: 'Dogs. — The total number of dogs registered in Sydney is only 1766; there is reason to believe that the number prowling about the streets without any ostensible owner is upwards of 3000.'

Over 3000 urban feral dogs in the relatively small town of Sydney was an astronomical figure! The threat to community health and safety would have been immense.

* * *

The colony's domestic dogs at least had *some* fans. The Eora quickly saw the potential in their greater faithfulness and better hunting and guarding abilities. A deadly quarrel among some of the Eora recorded by David Collins confirms the popularity of dogs with the country's original inhabitants:

> Being themselves sensible of the danger they [some Eora people] ran in the night, they eagerly besought us to give them puppies of our spaniel and terrier breeds; which we did; and not a family was without one or more of these little watch-dogs, which they considered as invaluable guardians during the night; and were pleased when they found them readily devour the only regular food they had to give them, fish.[8]

So began the love affair that endures to this day. But while the dog became a genuine, permanent companion for the Eora in a way the dingo had never been, it also proved a liability. It was more dependent, didn't run off at sexual maturity, required more food, and bred twice a year. That meant twice as many dependents when the Eora were beginning to struggle to feed themselves.

Dogs will be dogs. The Eora's dingoes would have followed their people into the colonial settlement and mixed with the local dogs. There would have been a lot of fraternising between the two types, resulting in litters of crossbred types, born to both the Eora and the settlers.

Before colonisation, as we've seen, a maturing pet dingo owned by the Eora would drift back into the local wild pack. These wild packs, their natural prey eliminated or displaced, were now also partly preying on food sources introduced by the colonists.

* * *

Feral dogs and dingoes instinctively know that people mean food. Both would prove to be nothing but trouble for the British as their settlement drive continued.

If the colonists east of the Great Dividing Range thought they had dingo problems, they would have needed a long draught of laudanum and a good lie down if they could have seen what awaited on the other side of the Blue Mountains.

Meanwhile, in Van Diemen's Land where there were no dingoes, a far more bizarre and serious assault on the young community and the fragile environment had been raging. And it was solely the work of kangaroo dogs gone bad, caused by an inexperienced government desperate to avoid a repeat of the Sydney colony's food shortage.

CHAPTER 5

Three Dogs Conquer Van Diemen's Land

Just after the dog diverged from the wolf, the rising seas of the dying Ice Age filled a giant moat in southeast Australia, creating a large island – known to the Nuenonne people of Bruny Island as 'Trowenna' and to us as Tasmania. The rising sea waters also obliterated much of the complicated sea and land route from Southeast Asia to northern Australia that the Aboriginal Australians had discovered 48,000 years before. And so, protected by the sea, the Australian mainland and Trowenna remained in blissful, dingo-less isolation.

Five thousand years ago, the Asian dog contrived to cross the northern seas with Lapita mariners, and the dingo conquered the mainland. But crossing the southern moat proved an impossible nut for the dingo to crack.

In the end, it wasn't the dingo that Trowenna needed to worry about. Once the age of sail lured Europeans into the Pacific Ocean, it was only a matter of time before Trowenna suffered invasion. And it was not the dingo but the colonial hound, the kangaroo dog, that ripped the Trowennan idyll apart.

No chapter in the Australian domestic dog's history is as bloody as the kangaroo dog's wild rampage in Van Diemen's Land.

* * *

Dutch explorer Abel Tasman was a fellow determined to impress the people who mattered. In 1642, Tasman sailed to the remote west coast of Trowenna, then, heading south, landed at a place now called Blackmans Bay.

Further north, Tasman planted the Dutch East India Company Flag and claimed the land for the Company, naming it Anthony Van Diemen's Land in honour of his boss, the Governor-General of the Dutch East Indies. Mariners more pragmatic than Abel Tasman decided that 'Anthony Van Diemen's Land' was a bit of a mouthful, and mercifully abbreviated it to the slightly more manageable Van Diemen's Land. Tasman would have saved everyone a whole lot of trouble and confusion if he had just named his discovery after himself.

Van Diemen's Land was originally thought to be part of mainland New Holland, as all of Australia was then known, until James Cook named the eastern half 'New South Wales' in 1770. Up until 1798, no one knew about the moat, even though thirty-six convict vessels, and a few other ships, had sailed north right past it. No one aboard apparently bothered to take a decent look out to the port side.

It was ten years after the establishment of the Port Jackson colony that Matthew Flinders and George Bass established that a wide expanse of water separated Van Diemen's Land from mainland Australia. They named it Bass's Strait, which doesn't sound nearly as romantic as 'The Moat'.

The discovery of Bass Strait spared ships travelling to New South Wales via the Cape of Good Hope or India from having to round the bottom of Van Diemen's Land before heading north. It could save weeks of battling the Roaring Forties – though the shallow, obstacle-strewn waters of Bass Strait themselves posed a constant danger, and throughout the years proved to be a graveyard for unwary mariners.

* * *

The kangaroo dog dominated Van Diemen's Land's early history, and it ended up there because of the never-ending jealousy and distrust between the British and their perpetual enemy, the French. After a year of the uneasy Peace of Amiens in 1802, Britain once again declared war on France because Napoleon had ignored the terms of the treaty and was cheating on them, and no one likes a cheat. The government in London and successive administrations in New South Wales also suspected the French of having designs on New South Wales, Van Diemen's Land and possibly New Holland, the name which, by then, only applied to the western side of the continent.

It started innocently enough in 1772 with the Du Fresne expedition, and then in 1788, when the two ships of the La Pérouse expedition to the Pacific Rim sailed into Botany Bay on the very day the First Fleet departed that barren place for the better-resourced Port Jackson, a dozen miles to the north. La Pérouse was on a keeping-up-with-the-Joneses quest to exceed Cook's remarkable discoveries. He remained in Botany Bay for six weeks, refitting his battered ships, and recovering from a massacre of some of his crew that had occurred in Samoa.

When the two ships took their leave of New South Wales they disappeared. It is now known that they struck a reef and

foundered off Vanikoro Island in the Santa Cruz group, southeast of the Solomon Islands.

The next cause for suspicion occurred in 1793, when a second expedition was sent to search for La Pérouse, led by Antoine Bruni d'Entrecasteaux. D'Entrecasteaux spent over a month creating charts of Van Diemen's Land and snooping around Bruny Island and the d'Entrecasteaux Channel. He was kept busy naming places after himself, though he couldn't hold a candle to the future governor of New South Wales, Lachlan Macquarie, in perpetuating his name. With typical Gallic gall, d'Entrecasteaux did not swing by Port Jackson to pay his compliments to Governor Phillip on his way out, which raised disdainful eyebrows and further suspicions.

It was during the Peace of Amiens that Matthew Flinders, still doggedly charting the coast of Australia in the *Investigator* – with his cat, for shame – came upon a ship flying French colours south of the present-day site of Adelaide. The French ship was the *Géographe*, commanded by Nicolas Baudin. She and her consort, the *Naturaliste*, had become separated during a recent storm. Baudin and Flinders were both unaware that peace had broken out.

Flinders, ever cautious and suspicious, had the *Investigator* cleared for action and her little pop-gun cannons run out as he approached the *Géographe* – the flapping Tricolore ensign on the French ship's stern being an extremely poor advertisement of the intentions of the persons therein. Throughout the hostilities, the British and French scientific communities – who thought the war was stupid – had regularly exchanged information about their discoveries. Baudin also carried a Letter of Good Conduct from King George, protecting the *Géographe* and her crew from British military aggression. Obviously, such protection depended on British man-o-war captains asking questions *before* running out their guns and firing broadsides.

Napoleon had allegedly dispatched the expedition to verify whether the west and east of the continent were divided: a likely story. The French were mapping the southern Australian coastline, and Baudin had already spent an appreciable amount of time in Van Diemen's Land with the eminent scientist François Péron. Over several respectful weeks, Péron had established excellent relations with the Aboriginal people there, conducting anthropological studies. Baudin had provocatively named the southern mainland Terre Napoléon (Napoleon Land), in honour of the French Emperor. That would have really impressed Flinders, who came up with the name Australia on the same voyage.

The *Geographe* was in as bad a state as her scurvied crew, and after Flinders and Baudin stiffly exchanged charts and notes, the *Geographe* took her leave of the *Investigator*. She limped her way to resource-strapped Port Jackson to refit, and to allow her crew to recuperate. This was a burden for the colony that must have thrilled Governor King no end, particularly as the *Naturaliste* arrived not long afterwards in the same state of dire need.

While in Sydney, Baudin turned out to be a pretty good bloke – for a Frenchman. He got along famously with Governor King and couldn't wait to show him his magnifique charts of Terre Napoléon with French place names plastered all over them. That was the last straw for the governor. King liked Baudin (Flinders didn't – Baudin criticised his charts and hurt his feelings), but the governor suspected that the French planned to colonise Van Diemen's Land and the southern part of the mainland.

King wrote of his suspicions to Baudin, who vehemently denied the accusation in both official and personal letters. However, Baudin's pet monkey had also just died, and he needed to help himself through the grieving process. So, the Frenchman got on the sauce with Lieutenant-Colonel William Paterson (recovering from being shot by Macarthur) and allegedly confided to Paterson that Van Diemen's Land might make a nice French colony.[1] Loose lips sink ships, *mon capitaine!*

Everyone in Europe was learning the hard way what kind of neighbour Boney was. Governor King had no intention of sharing any more space in the southern hemisphere with the Corsican despot, under any circumstances. King decided to establish a British presence in Van Diemen's Land forthwith, if not sooner. In September of 1803, he hurried off the newly arrived Lieutenant John Bowen to command a small penal settlement at Risdon Cove on the Derwent River, at the southern end of the island.

King then put a bomb under the Colonial Office in London regarding the overt French interest in nicking southern Australia. London promptly dispatched the convict ship *Calcutta* and an armed supply vessel, the *Ocean*, to stymie Boney's outrageous antipodean ambitions. Their mission was to start a second new penal settlement at Port Phillip Bay, on the southern tip of the mainland.

The man in charge was David Collins, the former judge-advocate of Governor Phillip's fledgling colony. Having served his tenure in New South Wales, Collins had returned to England in 1796, but he was sent back in 1803 to serve as lieutenant-governor of the southern outpost.

Collins chose a sandy, infertile part of Port Phillip Bay, which proved to be spectacularly unsuitable. The natives were hostile and getting hostiler, and even though Collins's previous experiences had seen the *Calcutta* and *Ocean* bring several suitable hounds with them, fresh meat was near-impossible to obtain. The only quadrupeds seen were dingoes and a couple of unobtainable kangaroos which had the good sense not to get too close to the smelly British and their dogs. The fish weren't biting; crayfish was the only easily accessible seafood, but even crayfish gets monotonous when it's the only fresh thing to eat. The people succumbed to scurvy and other diseases. The marines grumbled and were on the point of mutiny.

Faced with deteriorating health and morale and the real possibility of failure, Collins made an executive decision and, in February of 1804, after receiving approval from Governor King in Sydney, had the canvas outpost pulled down and packed up, and set sail across Bass Strait and down the east coast of Van Diemen's Land, heading for the Derwent.

Later in the year, Governor King dispatched Lieutenant-Colonel William Paterson to set up a colony at Port Dalrymple on the Tamar River in the island's north, as a means of securing Bass Strait. Paterson found the first site unsuitable. After another relocation, Paterson settled on a site further up the Tamar at the confluence of the North and South Esk Rivers, the site of present-day Launceston.

As the years passed, there would no longer be any need for the Van Diemen's Land colonies to pretend they were bulwarks against a possible French land grab. The French had done with their half-hearted snooping; it had been more than enough for them to learn the lay of the land and send the British into a flap. Crippled by war, and the ultimate defeat of Napoleon at Waterloo in 1815, they reached the conclusion that the French

stealing Van Diemen's Land from the British, who stole it from the Indigenous inhabitants and the Dutch (who tried to steal it from the Aboriginal people, but never got around to it) was never going to happen.

* * *

The dog's story since it arrived in Australia is truly unique, but nothing compares with the kangaroo dog's influence on the lives of both the first people and the colonists of Van Diemen's Land, and indeed on the environment itself.

The British invasion of Van Diemen's Land was the beginning of the end of ancient Trowenna. Yet in the early years, as in New South Wales, the narrow-visioned invaders found themselves hopelessly out of their depth in an environment that hadn't evolved to support voracious thieves.

Trowenna had a finely balanced ecosystem, and a small human population who lived in harmony with their pristine environment. Its carnivorous marsupial predators were the cat-like quolls, the stocky Tasmanian devil, and the vulnerable thylacine, or Tasmanian tiger.

The thylacine was an animal that was never going to impress the British invaders. A medium-dog-sized carnivore with an astoundingly wide gape and stripes across its back, a stiff, kangaroo-like tail and its secretive habits made it easy for the insecure and ignorant invaders to distrust.

The colonists rarely saw the thylacine in the early days of Hobart Town and Port Dalrymple. Its bad publicity really began when people ventured into the Midlands.

The thylacine would acquire a multitude of different names from a foreign people who were never going to look upon it with anything approaching understanding. Many of them called it a tiger – the Van Diemen's Land tiger, the bulldog tiger of the greyhound tiger. For an entirely unfathomable reason many called it a panther. Others called it a wolf – the marsupial wolf, the striped wolf, and remarkably, the tiger wolf, even though it is neither feline nor canine. But wait, there's more, and the names become even more derogatory: the zebra opossum, the dog-faced opossum, and ridiculously, the opossum hyena, the native hyena, and even more preposterously, the hyena tiger.[2] Bearing in mind the old-world opinion of wolves, tigers and hyenas, the names given to the thylacine confirm how little the Vandemonian Britons thought of it.

It is incredible to think that prior to 1803, neither the fauna nor the humans of Trowenna had ever seen a dog, and only a handful who met Péron or passing sea farers had ever laid their unlucky eyes on a European other than the rapacious sealers who frequented the Bass Strait islands. The moat had left Trowenna an isolated island paradise for thousands of years. But nothing could protect the island from the adversarial British attitude towards the fauna, the Aboriginal inhabitants and the environment.

* * *

Shortly after the establishment of the tiny Risdon Cove settlement, Collins and his two ships arrived from Port Phillip Bay. Collins assumed command and moved the settlement across the Derwent to Sullivan's Cove, the site of today's Hobart. Back then it was known as Hobart Town.

The over 400 people who came with Collins put a severe strain on food stocks, and resupply was nonexistent. The

Napoleonic Wars were distracting Britain and slowing the growth of her dependent dominions. With the umbilical cord damaged, the far-flung southern outposts had to fend for themselves, and they weren't coping well. The Colonial Office in London expected Sydney to keep Van Diemen's Land supplied, but for some strange reason, very few of those supplies ever made it south.

The Hobart colonists, missing their full ration of traditional staples of rock-hard salt beef and pork, bread, flour, tea, sugar, rum and wine, thought they were starving. Things never got anywhere near as bad as they had in Sydney, but Lieutenant-Governor Collins had experienced those bleak years of near-starvation first-hand. He took pre-emptive measures to ensure the new colony would not suffer famine.

Sydney had escaped outright starvation – just – and it had been largely thanks to the most popular animal in New South Wales, the kangaroo dog. Officers and other prominent members of society had brought the big hounds to Van Diemen's Land for sport, and as insurance against hunger. There was a ready supply of kangaroos and emus all around Hobart, and Collins, maintaining the English law that permitted only the upper classes to hunt and so ensuring convicts would remain reliant on the kindness of the Lieutenant-Governor, sent the officers out with kangaroo dogs to harvest the local wildlife.

That decision was a declaration of war against the native fauna, and in consequence, against the Aboriginal Tasmanians, the Palawa. The kangaroo dog, the saviour of New South Wales, became an offensive weapon in Van Diemen's Land.

Without the kangaroo dog's deadly efficiency as a hunter there would have been no Van Diemen's Land as we know it, and it's hard not to think that might have been a good thing. Regrettably, in Van Diemen's Land, the sweet-natured, sentimentally inclined kangaroo dog became an unstoppable force of destruction.

These kangaroo dogs were the product of several generations of hard-core kangaroo hunters. They knew their business, and were fearless killing machines with a blood lust that matched even their own lack of self-preservation.

Port Dalrymple (at the mouth of the River Tamar, downstream from present-day Launceston) suffered the same hardships as Hobart. Such was the perceived need for supplementary protein that a market was established for kangaroo and emu meat. Government stores set up in both colonies were hubs for the receipt and distribution of game meat. The neglected colonists did well on the lean, healthy meat, and clothing and footwear were made from the durable hides – the strongest hide for its thickness, being five times stronger than sheep and goat hide and ten times stronger than cow hide.

The officers found kangarooing a lucrative pursuit because game was everywhere, and the government, which had an obligation to provide every person with rations, paid well.

Convicted pickpocket-turned-author George Barrington writes that in the early days, the government stores were filled with the game killed by just four 'gentleman' hunters:

> As a proof of the utility of the kangaroo in this infant
> settlement, it seems that, owing to a recent temporary
> disappointment in receiving the Government supplies, the
> hind quarters of that animal were received into his
> majesty's store, at sixpence per pound, to victual the troops

and convicts, in the proportion of 7 lbs. for seven of salt
beef, or 4 lbs. of pork. Within the course of six months,
upwards of 15,000 lbs. weight had been tendered, although
only four gentlemen had hunted for the purpose.[3]

The kangaroo dogs were able to kill so many kangaroos and
emus that starvation no longer threatened, and everyone ate
more protein than they ever had. Except the local Palawa.

The inevitable over-exploitation decimated local game
populations, and those that survived the initial slaughter made
off with great haste for safer country. This meant the officers
had to venture deeper into the Midlands, where the
unimpressed Palawa waited in the shadows with poised spears.

This unpleasant prospect had the officers thinking that
perhaps the laws preventing commoners from hunting might
be a silly idea after all. So the officers and other kangaroo-dog
owners began to send their servants, convicts all, out into the
Midlands with muskets and dogs to hunt for meat, and crossed
their fingers as they waved them goodbye.

It was a decision that would haunt Van Diemen's Land for a
long, long time. The kangaroo dogs would become the
convicts' ticket to independence and freedom.

Vandemonian Britons became addicted to kangaroo meat,
and to the kangaroo dog with the Midas touch. Supply of
kangaroo dogs could never meet demand, there being a high
injury and mortality rate among them.

'I have a pack of kangaroo dogs as good as any in the whole
country,' big-noted the Deputy Surveyor General, George
Harris, in 1805, 'namely Lagger, Weasel, Lion, Boatswain,
Brindle, and with those dogs I scarcely ever go out or send out
(for I have two huntsmen) but get 3, 4, 5, or sometimes 8

kangaroos in a day or two – Some of the kangaroo stand 6 feet high and weigh from 100 to 130 or 150 lbs and fight the dogs so desperately so as sometimes to kill them and very frequently to wound them sadly.'[4]

Harris also hunted emus with his dogs. They were coursed in the same way as kangaroos but were just as dangerous, if not more so, able to break a dog's bones or disembowel it with a kick of their powerful legs. They also moved much faster than kangaroos, and only the swiftest kangaroo dogs were capable of catching them. The emu, however, stuck to clear, open country, whereas the kangaroo headed straight for the nearest patch of scrub.

The kangaroo dog became the most expensive domestic animal in Van Diemen's Land, and probably the most expensive dog in the world at the time. At the height of the madness some unweaned puppies fetched £80. That sum equates to many thousands of dollars in today's money.

Kangaroo dogs supplied the British settlers so well that even the convicts in Van Diemen's Land were better fed, and in some cases enjoyed greater liberties, than most Britons anywhere in the empire, including England. And it was all the doing of the kangaroo dog and a desperate, fearful ruling class with no foresight.

Kangaroo dogs made fortunes for the lucky owners whose dogs didn't disappear into the bush with their convict handlers. Dog theft was rife, and a crime that would earn the thief a couple of hundred lashes with the cat-o'-nine-tails, or a hanging – if he got lucky.

A great many of the convict kangarooers faithfully served their masters, enduring the dangers and hardship of the bush and keeping the government stores well stocked. But there emerged a dangerous sub-class, armed and roaming free over the Midlands. Many of these liberated convicts discovered that doing as they pleased in the bush was living the great Vandemonian dream, and they had no intention of returning to the colonies and slavery. These men quickly established the blackest of reputations, and a black market in kangaroo meat.

These outlaws bore Brown Bess muskets, which, being smooth-bored, fired a round lead ball. The Brown Bess was highly inaccurate, and worse than useless for hunting. But in the hands of a lawless man it proved a pretty handy tool for armed robbery, kidnap, rape and murder of Aboriginal people, uncooperative settlers and fractious competitors.

The Brown Bess was a short-range man killer, and little else. It was the kangaroo dog, not the Brown Bess, that sustained this dangerous sub-class. A convict in the bush with a pack of kangaroo dogs was king. Brown Bess just made him a nastier, more influential one.

The bushranging kangarooers became skilled hunters, but they were totally dependent on their dogs for survival. While tempting because of their superior size, the big grey kangaroo bucks were just too dangerous for these valuable dogs to tackle. The bushrangers needed to condition their dogs to pursue and kill the smaller, but faster, does and juvenile bucks.

The injury and mortality rates among these dogs, living rough and working tough, would have been high. There was almost no medical care available. Seriously injured dogs would have been dispatched on the spot. So when the local bushranger and his Brown Bess came calling, looking for new dogs, the owners were probably more than happy to oblige.

Because bushrangers were unable to sell their meat directly to the government stores, they sold their harvest to legitimate hunters, or to the tentative settlers who had established small farms in the Midlands in the wake of the kangarooers. These middlemen often supplied the bushrangers with gunpowder and round shot, and on-sold the meat and hides to the government. Some bushrangers would have bred up their dogs in the bush, and sold puppies to settlers and other hunters whom they hadn't yet robbed, raped or murdered.

It is highly unlikely the bushrangers sourced dogs from the Palawa, unless they took them from newly deceased owners, but not all bushrangers were tarred with the same brush. Some had good relations with the local Aboriginal people, but often the relations between the bushrangers and Indigenous people ranged from murderous to homicidal. Kangaroo dogs gave the bushrangers a huge hunting advantage over the Palawa armed only with spears and clubs – but the Palawa were no fools and would soon change their tactics.

In April 1807, the governor of the northern colony, William Paterson, he who had dobbed on Baudin for eyeing off Van Diemen's Land, wrote to the British Government to complain about the practice of allowing convicts to take firearms and kangaroo dogs into the bush in the pursuit of meat. He argued that the colony had become kangaroo-mad, that the number of bushrangers had exploded, and that it would be years before the lawlessness could be eliminated. His misgivings proved to be correct on every count, but he didn't put his hand up for having allowed the idiotic practice in the first place.

By 1814 the colony was self-sufficient and no longer needed kangaroo meat to survive, but there was still very high demand for kangaroo skins. The bushranger problem had become so acute that Lieutenant-Governor Thomas Davey offered an

amnesty for any bushrangers who were prepared to surrender
to the tender mercies of the regime. Very few bushrangers
took up the Lieutenant-Governor's generous, but 'too-good-
to-be-true' offer. So Davey proclaimed martial law, revoked
tickets-of-leave, set a strict curfew, and to strangle the
bushrangers' livelihood he banned the sale of kangaroo skins
and ordered all kangaroo dogs be shot on sight. Then he began
a relentless campaign to exterminate the bushrangers.[5] The
successful end of the kangarooing bushranger problem had the
unintended effect of leaving hundreds of kangaroo dogs to
their own devices in the bush.

* * *

The fauna of mainland Australia had had 5000 years to adjust
to the aggressive dingo. Kangaroos had long learned ways to
defend themselves against dingo attack, and the settlers'
kangaroo dogs in New South Wales had quickly learned
that kangaroos were to be respected. By contrast, the
Vandemonian kangaroos' only native threats were the reclusive
thylacine and the Palawa, who were usually only dangerous
within spear-throwing range. Yet the kangaroo and the emu,
when pursued by the far more efficient kangaroo dogs,
displayed the same instinctive defensive responses as their
mainland relations.

The kangaroos of Van Diemen's Land soon found
themselves preyed upon by a relentless predator much larger
and faster than the thylacine. Even though they were capable
of injuring or killing kangaroo dogs, the reality was that they
were only lethal once they had bailed up or were wrestling
with tentative dogs on the ground.

Dogs and wolves are far more confident during the chase, self-preservation forgotten the faster they run. They become understandably indecisive when face to face with a more confident, stationary opponent.

Around the Sydney Cove colony, the kangaroos were not as prolific as they were in the Midlands. The country of eastern New South Wales was scrubby, and making for dense cover was the kangaroo's most successful escape strategy. But in the lightly wooded, open country of the Midlands, though they often succeeded in killing individual tormentors, the kangaroos didn't stand a chance against the numbers of kangaroo dogs set against them.

Below is a summary by George Barrington of a lengthy letter written by an unnamed gentleman (probably Robert Knopwood, the Hobart-based pastor, magistrate and kangaroo-meat entrepreneur) to a friend in England. It is the most detailed account of kangaroo hunting in early Van Diemen's Land, and graphically describes the rapacity of the kangaroo dogs and the dangers of tackling kangaroos. Knopwood states that usually no fewer than three dogs at a time were set on a kangaroo, and most hunters kept at least seven or eight dogs. The dogs' habit of leading their master to the kill is astounding:

> After having run a considerable distance from view, they will return frequently after a lapse of two hours; and if they have killed, one chosen hound will intuitively conduct his keeper through an almost impassable brush, for miles, to the exact spot the reward of their vigour lies in, and if it was not for this quality of the dogs bred in this country, nine out of ten of the kangaroos killed would be lost.[6]

Dogs could not hunt more than three times a week, and many were miserably maimed or killed by kangaroos. Judging by Knopwood's description, it is surprising that many dogs survived at all:

> The chest of one was cut completely across, and so deep,
> that his lungs could be plainly seen, added to which, his
> right side was severely laid open by one kick. The other
> had the whole under part of his belly deeply gashed, the
> scrotum entirely cut away, and his mouth enlarged from
> nearly ear to ear! Yet these dogs, in five weeks, actually
> were again more savage, and as vigorous as ever![7]

It was common for the kangaroo dogs to wear out their paw pads to the point where they bled. They were quite often fitted with kangaroo-skin boots and long-necked collars to protect their throats from being ripped. The effectiveness of dog boots depended on the tolerance of the dog wearing them. They certainly provide cushioning for worn pads, and some dogs don't mind them, but others will chew them off and prefer to run on raw pads.

John West, the author of *The History of Tasmania* (1852), also wrote of the island's kangaroos ('foresters') and their battles with the kangaroo dogs:

> The Forester, the male being known by the name of
> 'boomer,' and the young female by that of 'flying doe,' is
> the largest and only truly gregarious species, – now nearly
> extinct in all the settled or occupied districts of the island,
> and rare everywhere. This species afforded the greatest
> sport and the best food to the early settlers, an individual

weighing 100 to 140 pounds. It is much to be regretted
that this noble animal is likely so soon to be exterminated.
Large powerful dogs usually hunted it, somewhat similar
to the Scotch deer hounds; and when closely pressed had
the remarkable peculiarity of always taking to the water
where practicable.[8]

Elsewhere West includes a regrettable description of a
kangaroo chased by a pack of English foxhounds, which
pursued a large buck for 14 miles before luck finally enabled
them to kill it. Kangaroo dogs usually ran down their prey
well within a mile or two.

There was a brief dalliance with the concept of riding to
kangaroos with English foxhounds in Van Diemen's Land, and
it is utterly remarkable that no foxes were imported for hunting
as they were on the mainland. Despite a couple of relatively
recent sighting hoaxes, Tasmania remains fox-free today.

Meanwhile, the emu stood no chance in the face of the
kangaroo–dog onslaught. It could disembowel a dog with its
deadly toenails, but kangaroo dogs were able leapers on the
run and could clamp onto a running emu's neck, killing it
instantly. Robert Knopwood used his kangaroo dogs in
coursing emus, which could be as dangerous as kangaroos if
the dogs were inexperienced or careless:

> This bird we catch here in great numbers, but it requires
> the fleetest of our dogs, which are frequently distanced:
> they weigh from 40 lbs. to 100 lbs. and will kill our
> strongest greyhounds by one blow of their talons. The
> most sure method the dogs have of killing the emu, is to
> seize them by the neck, in which they are extremely

sensible; and some of our old hounds are so well aware of
this, that they will often take two in one running.[9]

The collection of easily found emu eggs during the laying
season in May and June would have rapidly contributed to
their decimation. People and kangaroo dogs drove the
Tasmanian emu into rapid extinction, and by the mid-
nineteenth century it was no more.

* * *

Just like in New South Wales, the hungry invaders didn't give
a hoot what effect the decimation of native food resources had
on the Palawa.

For a people who had no experience of dogs, the native
Vandemonians, after an understandable period of
familiarisation, were quick to embrace the obvious advantages
of the kangaroo dog. Conversely, the Eora of New South
Wales, who had experienced 5000 frustrating years of part-
time dingo ownership, were a maritime people and did not so
heavily rely on kangaroo for their daily staple. Initially they
seemed to be more interested in obtaining little toy spaniels
and terriers.

Kangaroo dogs terrified the Palawa at first. They had no
conception that an animal could be so aggressive, so swift, or
capable of working so cooperatively among themselves and
with their handlers. The Palawa had much to learn, but they
learned it with astonishing speed, proving themselves to be
first-class handlers. After seeing kangaroo dogs swiftly running
down and killing game – something that might have taken
them hours of patient stalking without any guarantee of

success – the Aboriginal people soon realised the great benefits of hunting with dogs.

Certainly, they speared kangaroo dogs in early conflicts, but they soon changed that strategy and took to bullying outnumbered convict hunters into handing over their kangaroo kills. Some of these incidents turned deadly. Robert Knopwood shows little remorse in discussing the decline of the Palawa, given their predilection for these ambushes and attacks:

> I think, from our ravages, we gradually accomplish the effect of driving the natives from all the parts contiguous to any of our camps … They often spear our dogs, and attempt to pay us the same compliment, and not unfrequently will waddy our huntsmen, when they have not been inclined to part with what kangaroo they may have killed.[10]

Through trade and theft, and twice-yearly breeding, the Palawa wasted no time in accumulating large numbers of kangaroo dogs. And they needed to, because their game resources were being rapidly diminished and the dogs were becoming necessary to their survival. They quickly became competent dog handlers, and even more attached to their kangaroo dogs than the Vandemonian Britons.

There is not a lot involved in training sight hounds for the chase; it comes naturally to them. They have an incredibly high instinctive drive. More difficult, though, is teaching them how to work in cooperation with each other, and how not to get themselves killed.

Holding up small, wriggling puppies and allowing them to watch the adults on the job would have been enough to get

them started. The next step, when pups were mature enough, would have been letting them run with the adults in the chase. Start too soon, and there would have been a serious risk of permanent injury.

Working cooperatively with other dogs in the kill could only be learned by experience, initially on smaller, less dangerous game like wallabies. After that they'd have to learn on the job, on larger game.

It is extremely unlikely that a kangaroo dog in constant work ever lived to a ripe old age. Wild dogs and dingoes live about half as long as domestic dogs, and a Vandemonian kangaroo dog would not have been much different given its lifestyle and exposure to danger. The turnover of dogs would have been high, which partly accounts for the large numbers owned by the Palawa.

The Indigenous people would also have appreciated the sporting aspect of coursing with dogs. Coursing kangaroos and emus in the open country soon became as intoxicating for the local Aboriginal people as it was for the British.

The historian James Backhouse Walker recorded an instance of the Palawa's gratuitous handling of their dogs:

> In hunting, they destroyed the game recklessly, and could
> not be restrained from killing the kangaroo as while their
> dogs would run. On an adjoining island, where there were
> large numbers of wallaby, the blacks, in three or four
> hunting excursions, killed over a thousand head. By this
> kind of wholesale destruction, kangaroo, once abundant in
> the neighbourhood of the settlement, soon became
> extremely scarce.[11]

The kangaroo dog wasn't fussy about whom it killed for. It rapidly levelled the playing field for the Palawa, and they became serious competitors for the kangaroo and wallaby, which they understandably considered to be theirs. That didn't go down well with everyone else involved in the kangaroo meat and skin trade, who regarded marsupial game as the property of King George.

For the Palawa, the kangaroo dog became both a blessing and a curse. They were better able to source their staple meat and stay on the move, and their dogs also provided them with an early warning system and defence against attack. But having large numbers of big working dogs to feed contributed in no small way to the depletion of native stocks, which only increased British antagonism towards the native Vandemonians.

Then, once kangaroos and emus became scarce, the Palawa turned their dogs onto the settlers' sheep. Because the dingo was present in New South Wales, the sheep were confined at night in a (usually) vain attempt at protecting them from predation. There being no dingoes in Van Diemen's Land, the sheep were traditionally left unpenned of a night. That made them easy targets.

Sheep killing brought the Palawa into direct and deadly conflict with shepherds and settlers. They had given as good as they got in early tit-for-tat exchanges, and two decades later, despite their dramatically dwindling numbers, they were still waging a deadly guerrilla war of attrition, in an ineffective effort to rid their homeland of the invaders.

Competing for kangaroo meat was one thing, but waging war against His Majesty's subjects and their property and stock was quite another. The gloves came off in 1823, when a particularly ruthless form of bureaucratic bastardry marched

in, in the person of Lieutenant-Governor George Arthur. His rule led to the Black War and the rapid near-destruction of Tasmania's first people.[12]

* * *

It is futile to try to call off a sight hound in pursuit of prey, and many kangaroo dogs became lost when out hunting. The Van Diemen's Land kill-a-thon created daily opportunities for the increase of feral dog populations, and it wasn't just due to hunting dogs that went AWOL. As groups of Palawa were massacred, killed or captured, surrendered or died of smallpox, their kangaroo dogs disappeared into the bush and sought out their fellows. The same thing happened to the dogs of bushrangers who surrendered, were killed or were captured. Large feral packs formed and terrorised the Midlands.

As we've seen, there are no more troublesome animals than feral dogs, and they are never more dangerous than when acting in concert. The kangaroo dogs had learned every trick of the dirty trade from their owners, who had set them on native and domestic animals, and sometimes even other humans. They had become indiscriminate dogs with no taboos and few fears. It would be hard to imagine a more menacing sight of a night than a pack of large, powerful, feral kangaroo dogs out looking for opportunities.

Wild dogs usually avoid humans, however they often cautiously maintain a loose proximity to human habitation, exploiting the opportunities created by people.

Feral kangaroo-dog packs became the scourge of central and eastern Van Diemen's Land in the mid-1830s. Everyone was starting to think they could finally get on with things, but the Vandemonian colonists' chickens had come home to roost, and there was to be no peace any time soon.

The feral kangaroo-dogs caused serious localised problems for the developing wool and fat-lamb industries of the Midlands and the northwest. Farmers were losing hundreds of sheep a year to wild dogs. The place was a mess. Wild packs besieged small holdings and even haunted the colonies, threatening and attacking people.

Enough was enough. The interminable depredations were an embarrassment to Arthur and his successors, and a constant reminder of the failure of their predecessors to clean up the problem. The feral kangaroo dog had trodden on the most vindictive toes in Van Diemen's Land, and its fate was sealed, though in many parts of the island it was the thylacine that copped the blame for much of the feral-dog destruction.

From as early as 1814, successive administrations ordered the kangaroo dogs' destruction. The orders applied to all kangaroo-dogs, domestic and feral. They left the dirty work to kangaroo dog owners and the settlers. The feral kangaroo dog's life expectancy plummeted. The feral dogs were relentlessly shot, poisoned and trapped, and their dens and puppies destroyed (a practice called 'denning'), until at last the feral packs were decimated in the mid-1800s. Van Diemen's Land had washed its hands of its former canine saviour.

It was an ignominious end for the kangaroo dog, once the most valuable animal in the colony. Ironically, left to their own devices, the feral kangaroo dogs would not have survived long term. The dingo is the only canine that has been able to thrive in Australia unaided by man. Feral dog populations on the mainland only continue by assimilation into the dingo gene pool.

Unsurprisingly, when the ugly social and environmental consequences of the dog's rampage were finally swept under the carpet at Government House in Hobart, the administration just pretended, in true totalitarian fashion, that it never happened.

* * *

The British invasion of Trowenna destroyed the natural order and eliminated the native people, drove the emu to rapid extinction, and severely reduced the natural range of the forester kangaroo, driving them to extinction throughout the Midlands. If these atrocities weren't enough, George Arthur also turned Van Diemen's Land into a super-max prison, and the picturesque island utopia became home to two hidden places of terror that made Devil's Island look like Gilligan's Island.

Macquarie Harbour, on the bleak west coast, was Van Diemen's Land's first ultra-punishment prison camp. It was designed as a place of terror for convicts who re-offended while serving their sentences in the mainland penal settlements. After its closure in 1832 Port Arthur, southeast of Hobart on the Tasman Peninsula, became the new state-of-the-art government-sponsored hellhole.

For a convict to escape from Port Arthur he had to negotiate a narrow, sandy isthmus called Eaglehawk Neck, which separates the Tasman Peninsula from the northern Forestier Peninsula. Sentries were posted at the neck to keep guard, but some convicts still managed to escape, including the bushranger Martin Cash.

Then John Peyton Jones, a junior officer of the 63rd Regiment, devised a way to keep the convicts on their side of the line in the sand.

It occurred to me that the only way to prevent the escape
of prisoners from Port Arthur in consequence of the noise
occasioned by the continual roar of the sea breaking on the
beach and the peculiar formation of the land which
rendered sentries comparatively useless, was to establish a
line of lamps and dogs.[13]

Peyton Jones had a team of convicts clear and level a strip of
land across the isthmus and lay a deep bed of white cockle
shells to reflect light from lanterns. He then sourced nine
savage dogs and had them chained and kennelled in such a way
that no convict could pass between or around them.

The persistent myth is that it was mastiffs that guarded this
'dog line'. There is an ill-considered bronze sculpture at
Eaglehawk Neck commemorating the dog line – and it only
helps to perpetuate the myth. The sculpted snarling dog is an
attempt to replicate an English mastiff, though it is much
smaller, and looks more like a tail-docked dogue de Bordeaux,
the French mastiff. Mastiffs, large, heavy-set dogs, are ancient
breeds that originated in the Middle East or around the
Mediterranean, and there were mastiff types throughout the
old world.

There have been mastiffs in Britain for over 2000 years,
and the Romans considered them the largest and fiercest
mastiffs of all. How they reached Britain is not known, but in
almost every case where ancient dog breeds turn up in unusual
places in the old world, it is the Phoenicians, the ancient,
wide-ranging traders, who are nominated as responsible. The
Romans appointed an officer whose duty it was to select
suitable specimens and ship them to Rome to fight men and
wild beasts in the Colosseum.

The dog-line monument at Eaglehawk Neck. Everything about it looks authentic, other than the dog. Pity, that. (Courtesy of author)

Johannes Caius wrote of them in 1576: 'This kind of dog called a Mastiff or Ban Dog is vast, huge, stubborn, ugly, and eager, of a heavy and burthenous body, but of little swiftness, terrible, and frightful to behold, and fiercer than any other cur because they are appointed to watch and keep farm places and country cottages from thieves, robbers, spoilers, and night wanderers.'[14]

The Northumbrian naturalist and master engraver Thomas Bewick published *A General History of Quadrupeds* in 1790. He described English mastiffs as:

> much larger and stronger than the bull dog; its ears are more pendulous; its lips are large and loose; its aspect is sullen and grave, and its bark loud and terrific. The Ban Dog is a variety of this fierce tribe, not often to be seen at present. It is lighter, smaller, more active and vigilant than

the Mastiff, but not so powerful; its nose is smaller … its hair is rougher, and generally of a yellowish grey, streaked with shades of a black or brown colour.[15]

Now, that was the very dog Ensign Jones needed for Eaglehawk Neck! From Bewick's and several other descriptions, as well as a nineteenth-century illustration, it appears the guardians of Eaglehawk Neck were a local type of ban dog. The illustrated dogs are coarse-haired, ranging from light to dark brindle (tan to brown or black hair, with light contrasting streaks), and are shown 'housed' in wine casks that no English mastiff could even squeeze its bum into.

The British mastiff was, and still is, an enormous, smooth-coated dog with little tolerance for extremes of heat or cold, and being the dog of the estate, it lived inside by the fire when not at work. British mastiffs were the rich man's guard dog, because only the wealthy could afford to maintain them. They were true giants, with a kindly disposition, though they were all business when roused into action by robbers, common trespassers and the inevitable poachers.

Ban dogs were savage headcases, coarse localised types rather than a distinct breed; they may have originally been a cross between a mastiff and a staghound (the large, extinct ancestor of the foxhound). Another suggestion is that ban dogs were a mastiff crossed with the Eurasian shepherd dog, which still has a fearsome reputation. During the height of the bushranger depredations and the Palawa assaults, settlers kept large aggressive dogs for protection. Some writers of the time referred to them as mastiffs, but they were surely ban dogs, and seem to have been readily available in Van Diemen's Land.

Sir John Franklin the Governor of Van Diemen's Land and Lady Franklin visit the dog-line of Eaglehawk Neck in 1837. The dogs don't look particularly happy to see them. It appears the feeling was mutual. (Courtesy of State Library of Victoria)

The ban dog was the poor man's mastiff. Any crossing with a large aggressive dog would suffice. With a coarser coat and a hardier constitution than the mastiff, these dogs would have been better able to tolerate exposure to the elements.

Henry Melville, a prominent Vandemonian newspaper owner and journalist published a book, *The History of the Island of Van Diemen's Land from the Year 1824 to 1835 Inclusive*, that was highly critical of the tyrannical governorship of Sir George Arthur. The lieutenant-governor had provided Melville with plenty of material to work with over the years. The ban dogs of Eaglehawk Neck got a dishonourable mention as well:

> Those out of the way pretenders to dogship were actually
> rationed and borne on the Government's books, and
> rejoiced in such soubriquets as Caesar, Pompey, Ajax,

Achilles, Ugly Mug, Jowler, Tear'em and Muzzle'em. There were the black, the white, the brindle, the grey and the grisly, the rough and the smooth, the crop-eared and the lop-eared, the gaunt and the grim. Every four-footed, black-fanged individual among them would have taken first prize in his own class for ugliness and ferocity at any show.[16]

That wasn't much of a rap, but it does seem to be an authentic description.

Charles White wrote of the dog line in *Early Australian History: Convict Life in New South Wales and Van Diemen's Land* that 'the dogs were generally of a large rough breed, mongrels of the most promiscuous derivation, but powerful and ferocious'. Nineteenth-century writers, Mr White included, tended to embellish upon apocryphal stories about novel contrivances like the dog line. Anything other than directly quoting Mr White would be doing his nonsensical little anecdote a great disservice: 'One of the family, who was permitted to roam at large, amused himself sometimes, and kept his teeth and temper in practice by running into the shallows and fighting with the sharks: and he frequently succeeded in dragging them ashore.'[17]

Let us hope the shark-wrestling ban dog was never imprudent enough to grapple with Government Billy, the legendary man-eating shark reputed (by the guards, no doubt) to be in the administration's employ. The convicts believed Government Billy patrolled the waters of the isthmus, waiting to devour hopeful escapees swimming for it. Further details of Government Billy's famed devotion to duty have not come to hand. Nor are there any records of his victims, but he had the convicts conned.

Keeping the unfortunate dogs of Eaglehawk Neck permanently chained, unexercised, exposed to the weather and seas, and living in their own filth created nothing but disease, illness, neurosis and heightened aggression. To further discourage convicts' thoughts of escape, the soldiers made it known that they fed the dogs raw meat to make them more savage. Science (and common sense) have long known that feeding a dog raw meat does nothing but fill its stomach, but the rumour would have had the desired effect, because the ignorant and subjugated believe such things. Some of the poor dogs were stationed on platforms set above the water to prevent escape during low tides. It's little wonder they had a reputation for insanity.

Large dogs have relatively short lives. The life expectancy of the ban dogs of the dog line would have been proportionate to the deplorable conditions in which they were kept. Which were a lot better than some of the convicts.

There is no record that a convict ever successfully challenged the dog line, but in an act of ironic lunacy, an infantry sergeant with a death wish (and presumably heavily intoxicated) was severely mauled when he decided to test the dogs' vigilance by trying to sneak past them.[18] The most notable Eaglehawk Neck story of attempted escape and subterfuge was the 'what-could-possibly-go-wrong?' effort of convict and ideas-mastermind, Billy Hunt. Billy contrived to dress himself up in kangaroo skins (whether this included a functional pouch we do not know) and proceeded to hop – yes, hop – towards the dog line and the guard station, which happened to be manned by hungry, armed, marsupial-eating soldiers. Billy Hunt must have either had extremely long legs, very short arms and hopped very convincingly, or the guards who spotted him had very bad eyesight and had never seen a real kangaroo before. The musket balls flew thick and fast causing the 'kangaroo' to throw up his front legs in surrender and declare that it was only he, Billy Hunt, and not really a kangaroo. You don't say, Billy!

Vulgar Australian vernacular has never forgotten the name Billy Hunt.

The dog line guarded Eaglehawk Neck until Port Arthur closed its cell doors for the last time in 1877. What became of the ban dogs after that is not known, but they wouldn't have made attractive adoption candidates, and were probably shot.

Whether Government Billy received a golden handshake and retired from the convict-eating game at the end of his contract is also not recorded, nor is the fate of Billy Hunt's kangaroo outfit.

* * *

A very different breed of dog also became commonplace in Van Diemen's Land in the early nineteenth century – and one that would enjoy a happier fate. They were small, hard-bitten terriers of Scottish origin that found their way there with Scottish soldiers or settlers. They were typically tough Highland Scots, with a strong protective streak and a willingness to challenge any perceived enemy, no matter how large or menacing. They were an amalgam of possibly four courageous little Highland terriers that entered fox and badger burrows and cairns, and fought their larger quarry out of their boltholes.

The rough-coated terrier, as it was known, was a big, angry dog in a tiny frame. It was extremely suspicious and highly territorial, and proved to be a great asset for people still enmeshed in the wars waged by the disenfranchised Aboriginal people. English estates used small, alert terriers as early warning sentinels for larger, less alert guard dogs. The rough-coated terriers of the central Midlands also worked in tandem with ban dogs for settler protection against marauding bushrangers and Aboriginal Australians.

The rough-coated terriers had a high sense of duty and tackled anything trespassing upon their territory – and in the Midlands, there were plenty of trespassers. Every homestead would have kept fowls, and there is nothing like poultry to attract vermin and predators looking for an easy feed. But rats and mice were taking their lives into their own hands if they dared to set up shop when a rough-coated terrier was in residence. The tiger snakes and copperheads that are drawn to poultry also infuriated the bellicose little terriers, which quickly developed a reputation as efficient snake-killers that no other breed has ever equalled, or even wanted to equal.

The Midlands were home to four marsupial carnivores: the thylacine, the devil, the large spotted-tailed quoll, and the smaller eastern quoll. All of them developed unhealthy addictions to poultry and eggs. Standing up to any of that lot would have taken a big ticker – particularly being pitted against the spotted-tailed quolls, which had a reputation for fierceness. But the little rough-coated terrier took them all on.

The little terriers protected vegetable crops and fruit trees from wallabies and possums, and Tasmanians have long used little terriers to go to ground after wombats (which they still call badgers). The wombat looks cute, but is a surprisingly dangerous cannonball of an animal and has mauled and crushed many a terrier to death underground. It took a courageous little dog to tackle a wombat.

The rough-coated terrier stood only 10 inches (25.4 centimetres) at the shoulder and weighed a tad over 10 pounds (4.5 kilograms). It sported a rough steel-blue and tan coat, and it was no oil painting. It is widely believed the terrier's ancestry comprised the rough old Scotch terrier (the progenitor of today's Scottish terrier), the elongated Skye terrier, and the ridiculously grumpy, and grumpy-looking, Dandie Dinmont. It is highly likely the cairn terrier also contributed to the rough-coated terrier's later development. With that ancestry, it is no wonder the little rough-coated terrier had more attitude and less glamour than just about any terrier breed.

Later in the nineteenth century, the rough-coated terrier began to be exhibited at dog shows. Dog judge Walter Beilby, the author of *The Dog in Australasia*, quotes a contributor to a popular canine journal, who uncharitably described the rough-coated terrier as:

> an unmitigated mongrel, and only fit to use where snakes were too numerous to risk a dog of any value. People should be grateful that the Victorian Poultry and Dog Society has not allowed the name Australian to be prostituted to such vile uses and hung around the neck of a wretched mongrel. If whimsical or faddish people want an Australian breed, let them take up the dingo and try what they can make of improving him.[19]

The journalist in question was no doubt horrified when the 'unmitigated mongrel' received recognition as Australia's first breed in 1850. In 1897, its name was changed to the Australian terrier, and it has become one of the most popular dogs in Australia.

And with good reason. Pound for pound, the little Aussie terrier was unquestionably tougher and more courageous than any of the bigger, better-known colonial breeds that helped shape Australia. Most of the creatures it tackled were either larger or deadlier, and were quite often both.

No one knows how often the vigilance and gallantry of these dogs averted disaster for vulnerable pioneering families. Because of their diminutive size, their heroics were usually confined to the backyard, the homestead or the shed, and there is almost nothing other than reputation recorded of their deeds.

That a tiny mongrel from the Midlands has been so admired since the mid-nineteenth century, despite evolving under such adverse conditions, says everything about the pioneering little big dog that became Australia's first recognised breed.

* * *

Transportation to Van Diemen's Land ceased in 1853, and the island became 'formally known as Tasmania' the following year. Just twenty-four years after the final penal settlement, Port Arthur, closed in 1877, the Australian colonies would unite in federation and independence from Britain. The kangaroo-dog rampage, the attempted extermination of the Aboriginal people, and devices such as the dog line of Eaglehawk Neck were embarrassing anachronisms for colonies rapidly advancing towards the maturity of nationhood.

That advancement was in significant part due to the wool industry, and wool's growth was in no small part due to the development of the remarkable dogs that worked the nation's flocks. And it would be in the harsh interior of the continent that those working dogs came into being.

CHAPTER 6

The Kangaroo Dog, the Dingo and the Birth of Wool

Back across the moat in New South Wales, the kangaroo dog was still a jolly good fellow, and would be for another half-century. There, access to land anything like the Midlands only came twenty-five years after the establishment of the colony. In 1813, three leading colonial figures with extensive farming interests, Gregory Blaxland, William Lawson and William Charles Wentworth, worked their way across the hitherto impenetrable Blue Mountains, after things got a little cramped on the limited arable land squeezed between the mountains and the coast. Beyond lay endless expanses that would provide the foundation for Australia's future prosperity.

* * *

In the early nineteenth century, the most active man in New South Wales was the colony's easily offended multi-tasker Captain John Macarthur of the New South Wales Corps. As busy as a bee in a bottle, he was flat out as the CEO and majority stakeholder of the corrupt rum economy; in addition, he was a farmer in the American planter mould, an enthusiastic amateur lawyer and professional litigator, a committed dueller, a political intriguer, and a pioneer wool-grower.

Macarthur developed an unquenchable thirst for power, money and land. He was a big fish in a small, isolated pond half a world and half a year away from a war-distracted government in London. He developed a serious and potentially deadly set against anyone who obstructed his aggressive progress, particularly Royal Navy governors who threatened to become an irritating impediment to his corrupt business dealings. He daily challenged and undermined the authority of Governors Hunter and King. And in 1801 he duelled with, and shot, his boss Colonel Paterson. Paterson wasn't the first man who wanted to shoot Macarthur, and he wouldn't be the last.

Yet it is testament to Macarthur's highly developed time-management skills that with all that going on, he could still contribute in no small way to the establishment of the wool industry in New South Wales. Attending to the welfare and rehabilitation of convicts, however, does not appear to have featured high on his to-do list. There is only so much a man can do in a day. Being a law unto himself, accumulating enormous wealth, corrupting the economy and provoking powerful enemies seem to have been his priorities.

Putting a hole in his boss turned out a lot better for Macarthur, Australia, and indeed England, than even Macarthur could have imagined. Governor King sent him to London to face a court martial for his part in the duel and, not at all apprehensive of the consequences of his trial, Macarthur had the presence of mind to take some wool samples with him. The charges against him mysteriously disappeared en route to England, but amazingly his wool samples didn't. Miracles never cease! And a good thing, too, because British mills were desperate for wool, their supply from Spain having ceased during the Napoleonic wars. The expanding British empire was desperately short of good wool-growing land. And New South Wales was badly in need of a steady income.

Macarthur and other less wealthy, and arguably less unpleasant, pioneers (the 'flogging parson' the Reverend Samuel Marsden notwithstanding) had proven to themselves that New South Wales was an ideal place to raise fine wool-bearing merino sheep. Macarthur sought the opinion of wool merchants while in London and did a great job convincing them that the colony could deliver. Despite being on the other side of the world, New South Wales would capitalise on the strong demand for fine wool in Britain. Wool would prove to be ideal as an export commodity, because it was non-perishable, hard to damage and travelled well.

Macarthur had conjured up the perfect job for the colony – though ironically, the Macarthur family would not become a major long-term wool producer.[1]

The impressed merchants gave him the big thumbs-up – and the court martial never happened due to the lack of available evidence against him outside of New South Wales.

John Macarthur sailed for New South Wales in June 1805, smug and triumphant. He returned in his own ship, *Argo*, with no Argonauts alas, but with a ready market secured for his golden fleece and a few more merinos to add to his burgeoning flocks, some of which he had bought from King George. His greatest triumph, however, had been a land grant of 5000 acres of his choosing, wheedled from the government through the intercession of his powerful patron Lord Camden. Under the terms of the grant, Macarthur could claim another 5000 acres if his wool venture was successful.

Macarthur set about ramping up his political intrigues on his return to Sydney. He insisted that the 5000 acres he had inveigled from the British Government needed to be the Cowpastures, the best land found thus far in New South Wales. It was the land he had been desperately trying to get his hands on for the last ten years.

Governor John Hunter had declared the Cowpastures a no-go government reserve in 1795 for the safekeeping of the escaped Cape cattle, but Macarthur had other ideas. He decided he needed the Cowpastures more than the government did.

In his petition to Governor King, he included a proposal to buy all the government's wild Cape cattle. King refused him, putting the long-term welfare of the colony before John Macarthur's personal interests. However, in an attempt to keep the peace – and his own hide – he allowed Macarthur to take provisional possession of the Cowpastures while he took the matter up with the government in London.

* * *

Captain William Bligh, a naval protégé of James Cook, was a peerless navigator and foul-weather sailor, but when sailing easy or ashore he had all the tact and people skills of an enraged rhinoceros. Fletcher Christian and around half the crew of the *Bounty* deposed him in the infamous mutiny of 1789. Whenever he found himself at leisure Bligh turned his hand to hectoring and bullying his crew, especially Christian who for a time had been his particular friend. Seems they fell out over something, but what happened on *Bounty* stayed on *Bounty*. The crew had just spent months sampling Tahiti's delightful and accommodating attractions and most had taken up with Tahitian women who were shockingly liberally minded and most of the officers and seamen were, understandably, a little reluctant to leave. As soon as the *Bounty* departed Tahiti Bligh started 'topping it the tyrant' (in the seaman's jargon) and part of the crew, led by a conflicted Christian, mutinied. Bligh and eighteen loyalists were put over *Bounty*'s side in the ship's launch. With limited supplies Bligh navigated the small open boat from the island of Tofua in the South Pacific to Coupang in Timor in just over six weeks, a perilous journey of 3500 nautical miles. It still remains one of the greatest ever feats of survival and navigation – and holding your nerve. The court martial for the loss of his ship acquitted Bligh, but his behaviour remained open to question. He faced court martial again in 1804, this time over his 'tyrannical and oppressive and unofficerlike manner' while in command of the *Warrior*. The charges against him were part proven, and he was 'admonished in the future to be more correct in his language'.[2]

Bligh's career as a naval commander was all but over, but his patron from his *Bounty* days, Sir Joseph Banks, wrangled the foul-mouthed sailor the poisoned chalice of the governorship of New South Wales. Banks had an ulterior motive: he despised John Macarthur. So, Bligh succeeded Philip Gidley King as governor. His brief? To bring Macarthur and his Rum Corps cronies to heel and clean up the corrupt colony.

It was hardly an inspired appointment. In fact, it was downright idiocy.

Bligh, not unreasonably, saw Macarthur – by then a civilian, but no less influential – as the leader of the colony's immoral 'gangster empire',[3] and he had no intention of legitimising what he considered to be Macarthur's shameless land grab.

No one inflamed Macarthur quite like the even more combustible and hugely unpopular fourth governor of New South Wales. In the end, things got so ugly between them that Macarthur incited, then orchestrated, the New South Wales Corps mutiny known as the Rum Rebellion. Once again, Bligh was overthrown. Something, obviously not modesty, prevented Macarthur from claiming credit for the mutiny and the rumour that Bligh, a man of unquestioned courage, was found cowering under his bed when the mutineers stormed Government House.

Macarthur again fled New South Wales to exile in England as the result of a duel he fought with Colonel Joseph Fouveax (as it turned out, they both found the other to be an appealing target). He stayed in England for eight years, until 1817, avoiding any further prosecution for his involvement in not only the Fouveax duel, but also the Rum Rebellion.

Finally, Macarthur had obtained everything he would need to become a major wool producer in New South Wales. The only things the colony needed were a lot more sheep, and a lot more land. All the land they needed, and more, lay over the other side of the Great Dividing Range.

* * *

Governor Lachlan Macquarie, the fifth and most efficient of all the colonial governors, had been making up for lost time. He had ordered a road to be built over the Blue Mountains, and within only six months, a convict labour force of just thirty had hewn the road to riches, through 163 kilometres of

sandstone and clay wilderness, to the rolling timbered hills of what is now known as the central west of New South Wales. The year was 1815, and by anyone's estimation, it was an incredible feat. It earned those thirty convict men their tickets-of-leave and freedom.[4]

When they stood on Mount Victoria and gazed down on the golden west, even the soon-to-be-freed convict labourers could see the opportunity that stretched far beyond the shimmering distant horizon. It would be those wide-open spaces that would fast-track the development of Australia's wool and beef industries. From the humblest beginnings, these industries became the juggernauts that would be the making of modern Australia. And it would be the environmentally hardened native-bred working dogs that made that success possible.

But in 1815 the interior needed to be subjugated. And it was the kangaroo dogs that loped their way over the Blue Mountains, and laid the foundations of the colony's future prosperity.

* * *

Parties of wool and beef pioneers, some with bobtails, and maybe working collies, and all with kangaroo dogs, shepherded their stock along the Great Western Road, which wound its serpentine way through the conquered Blue Mountains towards the interior. Negotiating their stock safely to the Bathurst Plains was the first great challenge for pioneer sheep and cattle men venturing west, and the Great Western Road would prove to be the lifeline of the colony's wool and beef industries.

Governor Macquarie was all about establishing infrastructure. He chose the site for the colony's first inland town, at the dead end of the Great Western Road, and, flaunting tradition, he named it after someone other than himself. Having a nondescript patch of scrub named after him must have chuffed Lord Bathurst (the British Secretary of State for War and the Colonies) no end.

Bathurst – the nondescript patch of scrub – sat on the banks of the splendid Macquarie River, so named by the surveyor George Evans. The patch of nondescript scrub eventually became a town that served the settlers of the Bathurst Plains and beyond, to the splendid river Evans also named after Governor Macquarie, the Lachlan.

The Wiradjuri people had lived in the region west of the Blue Mountains for tens of millennia. They had somehow contrived to get it into their heads that their land was, well, theirs. To the British, that opinion was not only all too common, persistent and exasperating – it was also a brazen affront to King George, and inevitably led to bloody conflict.

Ironically, no one at the time understood the Indigenous Australians' distaste for the British better than the French. Naval captain Nicolas Baudin had had experience of Aboriginal people through his stay in Van Diemen's Land with anthropologist François Péron. In 1802, having recovered from the death of his monkey, he wrote and warned Governor King of the consequences of the British invasion. Baudin told King the Aboriginal people were hostile towards the colonists, and that punishing people who mistreated the locals was of no consequence to the Indigenous people. Baudin thought that the British had no hope of seeing the Aboriginal people happily cohabit with them – but that the British would soon

have peace anyway, because soon there would be no more Aboriginal people.[5]

Not every group of settlers just barged in. In 1835, at Port Phillip Bay in Victoria, John Batman tried to buy (read: dupe) the land from the local Aboriginal inhabitants, but Governor Bourke and his British masters vetoed the deal because it implied the Aboriginal people owned the land in the first instance. Which they did. Bourke asserted the Crown's ownership of all land in New South Wales, which gave no one, other than the Crown, the right to sell it.[6]

No matter what device settlers used to try to secure land, the Sydney how-to guide always ensured that things went pear-shaped for everyone when the dispossessed Aboriginal Australians decided enough was enough.

The British invasion of the Van Diemen's Land Midlands began with kangaroo hunters. In New South Wales stock went with the pioneers, and soon the word spread throughout Britain about the opportunities in New South Wales west of the Great Divide. Within a few years free settlers and their stock streamed across the mountains in such numbers that they quickly placed an unsustainable strain on the Wiradjuri and their limited natural resources.

The settlers' attitude to resource management in the interior was frugality with their own and profligacy with the Wiradjuri's. With no chance of regular resupply, the settlers conserved everything they brought with them, and fully exploited every natural resource. They built homes, furniture, sheds, barns, stockyards and enclosures for poultry, stock and vegetable gardens from local wood, bark, mud and stone. Traditional British staples like tea, sugar and flour had to last, and the settlers had only a finite supply of gunpowder and shot for their muskets.

The muskets would get plenty of use soon enough. The inevitable conflict with the Wiradjuri and the dingoes didn't take long to develop, because unless they had enough foundation stock to slaughter, the settlers' first choice for fresh meat, greenhide and leather was the kangaroo, the staple of the Wiradjuri and the dingo.

Before the coming of the British, these two age-old competitors had shared their resources amicably, in what was surely the most tolerant relationship between man and his competing apex predator anywhere in the world. There was usually enough kangaroo to go around, because hunting kangaroos was a time-consuming pursuit with no guarantee of success.

Of course, the kangaroo was a serious competitor for stock feed, and it didn't take long for the settlers and their kangaroo dogs to thin them out – so the Aboriginal people and dingoes turned their attention to the settlers' stock. Healthy cattle were large enough and cantankerous enough to be immune from dingo predation, but not Wiradjuri spears. Aboriginal people speared cattle for food and as an act of sabotage and revenge.

But for the dingo it was a different matter with the easy-to-kill sheep. The dingoes quickly became intoxicated by the harried sheep's wholesale, panic-stricken defencelessness.

In its natural state, the dingo only kills what it must to survive. The exception is the unnatural practice of surplus killing. Because when sheep are added to the mix, everything natural about the dingo goes out the window.

Surplus killing is human-like in its wantonness. Canids seem to be particularly prone to this excessive behaviour, which is an entirely wasteful predatory response: often, dingoes eat only part of one of their multiple victims. They also include

casual biting in their repertoire, which usually results in the sheep's death from infection. A slow, agonising end.

Some scientists think surplus killing is a predatory response triggered by fleeing prey (though foxes have also been found to kill stationary gulls roosting on beaches of a night).[7] Dingoes and dogs are programmed to attack large moving prey from behind. Stationary prey confounds a dingo and puts uncertainty in its mind. Likewise, the safest way of avoiding attack by an aggressive dog is to stand motionless.

Sheep are slow and uncertain, and mill together confusedly when threatened, with the vain hope of safety in numbers. It is their only frail defence. Fleeing invites attack, so when one of them takes its chances and breaks free it is as good as dead.

Most predators become inert soon after eating to satiation, but dingoes and dogs can kill a sheep, eat till they are glutted then continue chasing, catching and killing. A possible explanation for this atypical behaviour is that dingoes and dogs are susceptible to a serial-killer-like addiction to the adrenaline rush and excitement of killing. Domestic dogs can work themselves up into an exhilarated state and can nip, for example, when running about with overexcited children. The sheep's helpless milling response allows dingoes to attack them en masse, and that triggers an abnormal level of excitement, in which the biting and killing become an unnatural game.

Like all wild canines, dingoes learn hunting techniques from their parents. Once it started, sheep killing was taught to successive generations. Dingo predation on sheep is now perfectly natural, and there are plenty of dingoes that know no other way. They are as dependent on sheep for survival as the grazier. Every creature must choose the easiest way it can find to make a living. Sheep remove the choice. The dingo will not risk injury or death coursing kangaroos when sheep offer a safe, easy alternative.

The settlers snatched every last acre from the Aboriginal inhabitants, leaving them trespassers in their own country. But the conflict with the Wiradjuri escalated to the point that martial law was declared. With the rise of the squatters in the 1830s, powerful landholders, had the wherewithal to muster enough armed horsemen to take war to the desperate, fragmented Aboriginal people on the rangelands.

The settlers might have finally moved their human competitors on, but the dingo was a lot harder to shift.

The opening of the interior and the establishment of sheep-grazing land involved a lot of unnecessary tree and scrub clearing. Grasses did not grow well under the shade of the eucalypt and acacia forests, and a man was considered to be a bludger if he didn't remove almost every tree on his place.

Later in the nineteenth century teams of ring barkers, usually Chinese, worked their way through the grazing lands, killing millions of trees necessary for ecological balance. That practice has come back to haunt later generations with erosion and other environmental problems. Ghostly forests of still-standing dead trees, particularly the dense, iron-hard acacias such as gidgee and brigalow, can still be seen even today throughout Australia's marginal agricultural regions, evidence of the destructive zeal of the early settlers and the extraordinary durability of those timbers.

Land-clearing continued right up to the very edges of rough, unproductive or inaccessible country, or watercourses that provided refuge and corridors for dingoes. It created the perfect environment for the dingo to wage a guerrilla war against the sheep in open country.

Agriculturalism pulled the rug out from under the dingo. The settlers had supplanted the kangaroo and replaced it with sheep. With its natural and preferred prey diminishing under the weight of the kangaroo-dog onslaught, the dingo only did what any wild canine would do. From a behavioural standpoint, the dingo's response should have been entirely natural.

But in anyone's book, the level of predation was intoxicating and wanton. Seeing whole flocks of staggering, mauled sheep that needed to be put out of their misery was stressful and gut-wrenching for graziers. Mass sheep killings gave rise to the absolute hatred of dingoes and wild dogs on the front line of the wool industry. Graziers found themselves dealing with the reincarnation of the wolf of medieval horror stories.

The clashes with the dingo around Sydney, Parramatta and the Hawkesbury River had constituted little more than a phony war. It was only when the pastoral invasion of the interior commenced that the real dingo war began, and it rages unchecked today. Both sides have made gains and suffered losses, but there is still no clear winner.

Seemingly surrounded by the dingo, the early pastoralist conducted his daily economy under a constant state of siege. The first defence was shepherding during the day and guarding or penning small flocks at night. But even that didn't fully protect sheep from marauding dingoes addicted to sheep killing.

Like determined human enemies, the dingoes gave the settlers no respite. When they weren't killing of a night, their eerie howling and yowling, like the war cries of a raiding party, sent shivers through the lonely bark huts.

In the early days in the interior, boundary and paddock fences were nonexistent. Assigned convict labourers were vital to the success of the small pioneer holdings, and those men were often pressed into service as shepherds. Some were useful, but most, being men from the cities, were not. Shepherding in New South Wales was a much tougher gig than it was in England, and at times downright dangerous.

Even in a good year, grasses were sparse between the eucalypts that dotted the landscape. The shepherds had to walk flocks many miles a day to find enough feed for them. Isolated and vulnerable, shepherds and sheep became easy targets for attacks by the Wiradjuri, who developed a taste for mutton, lamb, and revenge. A shepherd camped out with a flock not only had to be alert to potential attacks by Aboriginal inhabitants – which was nerve-fraying and exhausting – but he also had to try to control nervy sheep. Many convict shepherds would have happily swapped their lot for life back in Sydney.

Kangaroo dogs served as both food-catchers and company for the shepherds. Some of the settlers would have used Scottish working collies as shepherd's dogs. Though game and loyal they weren't the sort of dog to act as a deterrent against dingoes and human intruders, but they would have served as effective watch dogs, being more alert and suspicious. More specialist sheepdogs would emerge later, but accompanying and working small quiet mobs of sheep with a pedestrian shepherd was not in itself particularly arduous work.

The Productions, Industry, and Resources of New South Wales was compiled in 1853 by co-authors Charles St Julian and Edward K Silvester.[8] They state that dingo hunting with large, savage kangaroo dogs was practised in the interior, where packs of two or three of the big hounds would run the animals

to ground. They were of the opinion, probably based on hearsay, that even one normal-sized dog was more than a match for a dingo.

Kangaroo dogs in New South Wales, while valuable, never fetched the ridiculous prices paid for them in Van Diemen's Land. Most settlers would have bred their own, and swapped dogs to ensure bloodlines did not become too close. Greyhound x deerhound was the favoured crossing for kangaroo work, but the much larger but slower Irish wolfhound was sometimes crossed with the greyhound, and these dogs hunted alongside men on horseback to run down and kill dingoes. Being huge, though, they were very difficult and expensive to keep.

* * *

By the mid-nineteenth century wool had become the unstoppable force in Australia, and no other environmental influence, other than drought could slow its progress, though the dingo has never stopped trying. But things were changing for both the dingo, and the kangaroo dog.

With the coming of fencing, sheep were no longer shepherded but were turned out into huge paddocks. Dingo predation increased, and as holdings grew, dog trappers were employed to purge the sheep lands. In some places, eliminating the dingo worked. In other places, it didn't. And everywhere it worked, it eventually created greater problems than it fixed.

Persistent hunting of kangaroos had reduced their numbers, and for a while it had looked like the kangaroo dogs would have little to do. Yet the arrival of fencing on the Australian mainland from the mid-nineteenth century coincided with advances in infrastructure, pasture improvement and water management. Graziers were enjoying a higher standard of living and had long ceased eating kangaroos, and with the Aboriginal people gone, and hundreds of thousands of acres of wheat and other crops planted, the kangaroos bounced back in no time at all.

Once again, the kangaroo dogs found a purpose in running down their old prey – though the killing now was simply wasteful culling, with more kangaroos destroyed than could ever satisfy the skin trade. Some kangaroo meat would have fed dogs, but the majority of culled kangaroos rotted where they fell.

The British obsession with eliminating Australia's native fauna did nothing but create problems for graziers. The kangaroo culling led to a proliferation of green blowflies come to breed in the putrefying carcases, and the resultant fly strike has been the bane of the wool industry ever since. The blowfly has maimed and killed more sheep than the dingo ever has. However, this wasteful practice has ironically provided easy food for a variety of native animals, particularly dingoes, and opportunistic birds such as crows, ravens, wedge-tailed eagles and a range of smaller eagles and kites. The smorgasbord of carrion that was the by-product of pastoralism also became a reliable food source for imported pests – foxes, and feral pigs, dogs and cats.

* * *

By the late nineteenth century, time and technology were starting to overtake the kangaroo dog. It still had a job, but that was about all. As a large-scale culling tool in an industry racing towards modernisation, the lagging kangaroo dogs had become ridiculously labour-intensive and ineffective. They represented a century-old method of kangaroo hunting that began as a response to the threat of starvation.

In the late eighteenth century, people hunting with kangaroo dogs for meat were prepared to be patient, given that their lives depended on success. They generally had close relationships with their dogs. The impatient graziers of the late 1800s needed their vermin destruction done as quickly and efficiently as possible, and relying on kangaroo dogs was becoming too hard.

The dogs were also an expensive liability: they had to be bred, then the puppies needed to be raised, while the adult dogs needed to be fed and securely kept. There was still an extremely high injury and mortality rate among dogs in work, and at least half a dozen working pairs of adult dogs and bitches were needed to ensure puppies could be produced, the wounded could recover, and at least half a dozen injury-free dogs were available for duty.

In the latter part of the nineteenth century, higher standards of living also gave Australians the opportunity to devote more time to companion pets. Dogs became a craze that has never abated. Australia is dog-mad Britain's child, after all. Imported breeds from around the world were enthusiastically embraced by a people bored with the same-old same-old.

By this stage the kangaroo dog had disappeared from urban areas, and had developed a regrettable, and undeserved, reputation for being aggressive – based on its feral rampages and the nature of its work. In fact kangaroo dogs made fine companions, but the people in the bush who owned them didn't want them as pets, and the folk who were capable of reinventing them as pets never saw them, because they only existed out in Woop Woop, and in old people's memories. The new pet owners were only interested in supporting the exotics and the newly created local working dogs.

So the kangaroo dog's popularity declined as its owners found it harder and harder to justify its existence. Redundancy was only a matter of time.

As soon as horses became readily available they became the primary source of transport, and a day's hunting in the wide-open spaces became a lot more demanding for kangaroo dogs. They were obliged to cover more country with their faster-moving masters, which was much more tiring than following a man on foot.

In addition, the introduction of barbed-wire fencing gave the kangaroos another escape tactic. Just as they had always headed for thick country when pursued, the kangaroos of the wide-open spaces of the interior, both reds and greys, would head straight for the fences and bound over them, ending the chase in most cases. The injuries suffered by kangaroo dogs running at full pelt into barbed-wire fences doesn't bear thinking about.

But it was the availability of repeating rifles and centre-fire and rim-fire ammunition in the late nineteenth century that finally rendered the kangaroo dog obsolete.

On the holdings where kangaroo dogs were still used, they were kept chained or caged well away from the homestead, with all the other farm tools; the fencing materials, ploughs and other horse-drawn agricultural implements. They no longer slept by the hearth or played with children. Unlike some of the 'harder', more independent scent-hound breeds, which live and hunt in large packs – foxhounds, beagles, harriers - sight hounds like the kangaroo dog are very reliant on human company.

Keeping kangaroo dogs in a commercial manner broke the human–dog pact, just as any form of neglect does. A kangaroo dog devoid of human attachment was a terrible creature in the making. It was only a matter of time before poorly managed kangaroo dogs escaped, took their own chances and turned wild. Feral dogs are a constant reminder of the failure of people to repay the dog's loyalty and dependence with adequate care and management.

It was in this guise, a feral menace, that the kangaroo dog would experience another resurgence in the 1930s. And it was all thanks to social and economic upheaval and a most unlikely little benefactor.

* * *

In 1859, another posh numbskull, a landowner named Thomas Austin had liberated twenty-four European rabbits at Barwon Park, a rural property near Winchelsea in Victoria. As Mr Austin watched their little white tails bobbing happily away, it is alleged he said, with no small air of confidence, 'The introduction of a few rabbits could do little harm and might provide a touch of home, in addition to a spot of hunting.'[9] Little harm indeed!

By 1886 the rabbits had reached Queensland. By 1900 rabbits were eating Western Australia and the southern Northern Territory out of house and home. By 1946 rabbits occupied over 4 million square kilometres of Australia, and their impact on agriculture and the broader environment was devastating.

People called the rabbit 'the poor man's mutton' – and the upside, if there really was one to the cute little environmental disaster, was that rabbits kept thousands of unemployed people fed in the bush and the cities, particularly during the Great Depressions of the 1890s and 1930s.

Australians became a nation of reluctant rabbit-eaters for several decades. But this time, hunters didn't turn to the kangaroo dog; it was the smaller, faster, less damaging greyhound and greyhound crosses and little whippets that were favoured as rabbiting dogs. Greyhounds enjoyed something of a renaissance during the decades of the plague. They had always had a following in Australia. Some people probably preferred them to kangaroo dogs, and they were popular for smaller, less dangerous game like the smaller wallabies and other little macropods. And greyhounds were ideally suited to rabbit work in open country.

It was during the Depression years of the 1930s that Australia experienced a sharp increase in sheep losses to feral kangaroo dogs. The kangaroo dogs were in an irreversible decline by the turn of the twentieth century, but they still had their adherents and they were still to be found on some properties and in the bush towns, but by then they were kept in smaller numbers purely for 'sporting' purposes.

Neglected and abandoned station kangaroo dogs that made their own way in the bush were joined by those abandoned by destitute townsfolk during the Depression years, and poorly trained dogs lost by inexperienced hunters. The millions of wild rabbits that devastated the rangelands sustained them when they couldn't find domestic stock to kill.

Enough kangaroo dogs were abandoned around the country during the Depression for them to make some serious trouble for a while. Outbreaks occurred right around the country at roughly the same time as rogue kangaroo dogs went on rampages, destroying hundreds of sheep in a few nights. With plagues of rabbits, they would never go hungry if sheep were not to be had.

At this point, in a repeat of the experience in Van Diemen's Land 100 years before, the kangaroo dog suffered a complete fall from grace on the mainland. And this time there was no going back.

There is no shortage of newspaper stories relating to the depredations of feral kangaroo dogs on stock during the 1930s. The addiction to mass killing affects all the canines, and Australian graziers suffered at the hands of not only the dingo, but domestic, semi-domestic (or semi-feral), and feral dogs. Graziers at Marulan in New South Wales were troubled by a wild dog whose attacks took a heavy toll on sheep in the district. It was finally caught in a dingo trap on Arthursleigh Station:

> The dog, which is said to resemble a cross between a
> kangaroo dog and collie, was brown with dark markings.
> It was 2 feet 8 inches high, and measured 7 feet from the
> nose to the tip of its tail. The tail was about a foot long.
> When found in the trap by J. R. Creswick, an employee of
> the station, the animal showed fight. Creswick had noticed
> the dog's tracks on the property, where it had killed
> 10 sheep in a few days.[10]

If these measurements are accurate, it would suggest the captured dog was an Irish wolfhound, or an Irish wolfhound x greyhound, one of the breeds that were popular for a time as dingo killers. Wolfhounds and deerhounds have the longest tails in dogdom, often hanging little over an inch (25.4 millimetres) from the ground. The wild Marulan dog must have lost most of its tail in an accident.

Wild kangaroo dogs even threatened and attacked humans. A young boy was lucky to escape with his life when two feral kangaroo dogs pursued him at Mount Morgan in central Queensland in January 1932. Workers coming into town from Limestone Creek, on the other side of the Mount Morgan mining company's big dam, heard distressed cries coming from the bush. On investigating, they found the boy in a state of panic, sitting on a bough high up in a tree.

The boy said he had been chased by two kangaroo dogs that had gone wild and were living in the scrub. To save himself he had had to climb the tree. It was only when they heard the rescuers coming that the dogs had slunk away into the bush. Another person travelling by horseback came by and said he had seen the dogs, which had menaced him, but he had been able to drive them off with his whip. Locals planned to shoot the dogs before they did more damage.[11]

The feral kangaroo dog issue was serious enough to be raised in the New South Wales Parliament. The Premier, Bertram Stevens, gave a noncommittal reply to Mr Aldill of the United Australia Party, who suggested a limitation of the breeding of kangaroo dogs, which, he said, were becoming a menace all over the State.[12] There was no government program put into place to eradicate the feral kangaroo dogs, but their days were numbered anyway. The Australian environment and the affected graziers would soon take care of that.

* * *

The kangaroo dog was the first dog of colonial Australia. For the first 100 years of colonisation, the big wire-haired hounds served their masters with selfless distinction. They saved Sydney, Hobart Town and Port Dalrymple from the threat of starvation; they accompanied and fed the explorers of the interior, and they did the dirty work in paving the way for pioneers no matter where they ventured. For over a century in New South Wales, Van Diemen's Land and the other colonies of Australia they were as common as Labradors and Staffordshire bull terriers are today.

The kangaroo dog was never going to win the hearts and minds of the average Australian, though it always had a few adherents. Generally it was nothing more than a tool – a large, expensive, difficult to keep and not very aesthetically appealing tool. Ultimately, the kangaroo dog was doomed. When it outlived its usefulness, it was not afforded the luxury of becoming a companion pet as so many other breeds were when they put their ugly pasts behind them. Its faithful heroics meant nothing, because like most tools, it was only valued for what it did, not what it was.

Kangaroo dogs were the ultimate canine paradox. They were sweet-natured, almost sentimental dogs created for the most brutal work against an innocuous-looking but dangerous herbivore in the harshest continent on earth. And they died in their thousands in the line of duty. The kangaroo dog's reward for its unflinching service to King, Queen and country was abandonment.

The kangaroo-dog paradox was no better exemplified than in its excessive sheep-killing behaviour when it was allowed to run wild. Fast, powerful, aggressive and indiscriminate, they caused localised problems way out of proportion to their numbers and were hunted down and exterminated with all the grim determination of people protecting their livelihood.

The demise of the feral kangaroo dogs on mainland Australia was inevitable, and by the mid-twentieth century the crisis was over. The dingo is the only dog type that can tolerate the Australian environment. The kangaroo dog was man's invention, not nature's, and their exaggerated long and lanky build was not up to the extremes of living wild in the various Australian environments in which they roamed at large. A dog living wild in the Australian bush and not a dingo type, won't be found in the same type in future generations. Its offspring will succumb to natural selection and eventually over several generations its progeny will become more dingo in type. Cosying-up to the local dingoes will move things along a little.

The mainland ferals interbred with the dingo, and for some time the extinct kangaroo dog's legacy of brindle and grizzled colouration would have been seen in the dingoes of the eastern ranges and other places where they interbred. In Van Diemen's Land back in the day there was no dingo for the feral kangaroo dogs to interbreed with, and through a combination of human intervention and natural selection they eventually disappeared leaving no living legacy of any kind.

Localised versions of the kangaroo dog were still occasionally found in outback Australia right up to the 1950s, but those remnants had been diluted with other breeds and were not the same type as the classic Scotch rough-haired greyhound x deerhound types.

The kangaroo dog is not usually included in the list of Australian-developed breeds. Yet despite its ugly job and ultimate, sad demise, it deserves recognition for its significant contribution to the establishment of the new order during Australia's toughest years.

But rural Australia had moved on by the early twentieth century, and the kangaroo dog was left far behind – forgotten in the wake of four truly Australian breeds that would revolutionise the beef and wool industries.

CHAPTER 7

The Beardies and Bobtails Find a Home

In Britain, all a stockman ever amounted to was a tenant farmer, no matter how talented and hard-working he might have been. In New South Wales, he could be his own man, with his own land, which cost him only courage and initiative. Once the Great Western Road opened the west, British stockmen and pastoral wannabes sailed to New South Wales as free men, the free selectors, looking for the opportunity to brave the hardship of the bush to select their own land and put down roots in the parched, thin soils. Some came with their own sheep and their own working dogs, merely transferring their pastoral concerns from the overcrowded United Kingdom to 'empty' New South Wales.

In the days when wolves were still common, shepherds kept two types of dog, a guard dog and a working dog, both confusingly referred to as shepherd's dogs. The first were dogs that provided protection for vulnerable flocks and lonely shepherds armed with nothing more than a staff and knife, particularly in the isolated hill country. A mastiff type,[1] these dogs varied from shepherd to shepherd and region to region. These large shepherd's dogs had a formidable reputation and

could more than account for themselves against marauding humans, wolves and feral dogs.

Once England's forests were cut down for farmland, the wolf had nowhere to hide, and by the turn of the sixteenth century it was extinct. The biggest threats to sheep were then the foxes, which could play havoc at lambing time, and the odd feral dog or starving peasant. These threats were obviously not serious enough for the shepherd to employ the guarding breeds, and the shepherd's dog used for guarding stock became consigned to history.

The shepherd's other shepherd's dog was the sheep worker. The English produced various sheep-working dogs, most were localised micro-types that are now extinct, but the English were not as heavily dependent on them as Australian stockmen would be, and perhaps they didn't have to be. Shepherding was the backbone of the wool industry in England for centuries. The need for farm workers was so great that at times experienced shepherds found themselves in great demand.

This changed following the union of England and Scotland in 1707, when Scottish shepherds, considered the best stockmen, found work on enclosed estates throughout England, and took their collies with them. The hardy, hard-working Scotsmen and their equally hard-working collies dominated the British pastoral scene for the next 100 years.[2]

The late eighteenth and early nineteenth centuries was a period of great change in Scotland's Highlands. Cheviot sheep from the English border area were taken to the Highlands to supplant the blackface sheep and cattle herds removed from crofts during the horrid Highland Clearances. English and Scottish shepherds from northern England were recruited to

THE SHEPHERD's DOG.

THE CUR DOG,

The shepherd's dog and cur dog from Thomas Bewick's *A General History of Quadrupeds* (1790). Of the shepherd's dog, Bewick said 'This useful animal, ever faithful to his charge, reigns at the head of the flock' and of the cur, 'They are larger, stronger, and fiercer ... the cattle have no defence against them.' Note that this is not the same 'cur dog' that caused so much trouble around Sydney. That was a nondescript mongrel.

the Highlands to manage the cheviots and with them came
their shepherd's dogs.

'Collie' is a broad term used to describe several breeds of
working sheepdogs developed by the Scots. There has been a
lot of fanciful guesswork employed in trying to determine
what 'collie' originally meant. Some insist 'colley' is the Gaelic
word for 'useful',[3] but Gaelic uses no word even remotely like
'colley' for 'useful'. That explanation is *not* useful.

Another theory is that the Scottish blackface Highland
sheep were called 'colley sheep', and that because the collies
had black faces too (they didn't), they were called 'colley dogs'.[4]
The speculation gets even more desperate: all the collies were
black − like coal (they weren't) − so they became known as
'coallies', which evolved into 'collies'. Who knows? But John
Jamieson's *Etymological Dictionary of the Scottish Language*, first
published in 1808, lists 'collie' as 'the vulgar name for a
shepherd's dog'.[5] Which is all we really need to know.

Collies are sheep-working dogs. Better-known breeds
range from the border collie and the bearded collie, to the
rough collie and its miniature version, the Shetland sheepdog.
Lesser-known and extinct collie breeds and types include the
Irish collie, the Welsh grey, and of course the entirely fictitious
fox collie, the dog reputed to be the (genetically impossible)
cross between a collie and a fox. Collies can be long- or shorthaired, and some, like the bearded collie, are very long coated.

The first working dogs to come to New South Wales in the
late eighteenth century appear to have been the bearded collie
and the legendary, enigmatic bobtail − sometimes known in
New South Wales (incorrectly) as the Smithfield.

* * *

The bearded collie was common throughout northern England and Scotland and had been since the seventeenth century. Traditionally known as the Highland collie, the Scottish bearded collie or the beardie, it was more a droving dog than a sheep-herding type. Some were born with tails and some without, and they came in all shades of grey, reddish fawn and black, commonly with white collie markings on their feet, tail tip, chest, with or without a forehead blaze which can join a whole or half collar. Such markings can be called tuxedo markings or Irish trim.

Such was the state and size of the early wool industry in New South Wales that men without dogs shepherded the small, valuable and vulnerable flocks. Space for sheep east of the Great Dividing Range was limited, and initially there was no pressing need for working dogs. Assigned convict labour was freely available, but as mentioned, most convicts made poor shepherds.

The bearded collie was one of the first sheep-working dogs to come to New South Wales, and on the small holdings around Sydney it managed. But once all the action moved west of the Great Dividing Range, the bearded collie faced rapid and permanent redundancy.

The soft, work-ambivalent beardies and bobtails, way out of their depth in continental Australia's brutal interior, would play no significant role in the growth of the mainland's wool industry. But beardies and bobtails imported into New South Wales in the early nineteenth century also found their way to Van Diemen's Land, which had a cooler, damper, more English-like climate. Settlers there discovered that they suited work on small holdings, and they eventually evolved into the dog now known as the Tasmanian Smithfield, a working

sheepdog and occasional cattle dog still popular on wool
concerns throughout the island state.

* * *

The Tasmanian Smithfield is squarish to rectangular in profile,
depending on the strain; its coat is shaggy and can be any colour,
though breeders favour bobtailed dogs bearing a solid colour and
white collie markings. The nose is black and the eyes are brown.
The head is in proportion to the body and the ears can be full
drop (floppy) or rose (small folded), but never full (erect) pricked.[6]

The Tasmanian Smithfield is not a breed registered with
the Australian National Kennel Council (ANKC). How such
an appealing dog dodged official breed recognition and the
grooming benches, scissors and blow dryers of the dog-show
scene is anyone's guess. Perhaps it's because the Smithfield
would look ridiculous all tizzed up, like a garbo in a dinner
suit. It is a rough and ready dog who looks best crudely
smartened up by the practical hands of shearers.

The Tasmanian Smithfield has developed a loyal following
in its home State. Because it is a genuine farm dog, the
overriding consideration is ability, so there is some variation in
appearance. Crossbreeding with border collies and kelpies is
common and the original type is being widely lost. The annual
agricultural show at Campbell Town in the Midlands stages
sheepdog trials and very casual conformation classes (beauty
competitions) with categories for Smithfields and Smithfield
crosses. The dog events are light-hearted and not affiliated
with any formal kennel club.

The Tasmanian Smithfield is an anomalous relic of a past
age. Like most of the collies, it was not cut out for the hard

interior of mainland Australia, but it found its niche in an Australian environment more like Britain, and has proven to be a valued, if not a brilliant worker there. It is a capable sheepdog and cattle worker, and being sociable and even-tempered, it makes an excellent companion. Its shaggy coat and drop ears mean it is perfectly suited to the freezing conditions of the island State.

* * *

The bobtail was a shaggy-coated drover's dog. It was probably the product of a cross between the beardie and the Welsh grey collie. An extra infusion of Scotch collie gave the bobtail better gathering ability, so it proved to be a handy worker, and sheep-friendly.[7] (These developments occurred when the original bobtail and the Scotch collie were working dogs, a long time before both breeds became ornamental powder-puff show dogs – respectively, the old English sheepdog of paint-commercial renown, and the rough collie made famous by Lassie.)

In Britain, the bobtail was just one of several diverse types that worked in and around the Smithfield livestock markets on the edge of London, and was more common south of the Thames River, while in the north the cur was the favoured drover's dog.[8] The bobtail that found its way to New South Wales was reputed to be a large, squarish dog, sometimes black, usually with a large white collar and ruff, and was preferred bobtailed.

When British free settlers began to arrive in New South Wales to take up 'vacant' land in the interior, their immediate need was droving dogs. The bobtails – the breed commonly

used around Sydney – were the obvious choice, because at the time there was no other.

Cattle were an entirely different proposition from small, vulnerable flocks of sheep. They were large, formidably armed beasts that were willing to confront threats as a united front. Their far-ranging habits and insular attitude precluded any kind of treatment that resembled shepherding, yet they became independent and pugnacious when left to their own devices in the bush. Mustering and management of wild and semi-wild cattle was a potentially dangerous undertaking that called for hard, forceful, courageous dogs capable of enormous stamina and endurance. The bobtail was developed in the cool, damp, British climate, working docile stock on small holdings and short-stage droving routes. Being a 'lowland' droving type, it was squarish in profile, and solidly constructed. It was not a lithe, loose-loined, lower-to-the-ground type suitable for working stock in difficult terrain. (The nimble border collie, a more recent development, is typical of the build needed for such work.)

Around the Hawkesbury River flats and the rolling downs of the Cowpastures, the bobtails coped. But the Great Western Road's tortuous route over the Great Dividing Range was a foreboding introduction to the conditions that awaited in the stinking-hot, bone-dry, flint-hard and flea-, tick- and burr-infested interior. The black bobtail, as it was sometimes known, was the pioneers' best cattle-moving option, but when it met the worst New South Wales had to throw at it, the bobtail would be found wanting.

No one would come to know that better than Jack Timmins the drover.

* * *

Back in Ireland, James (Jack) Timmins had been a member of the rebellious Society of United Irishmen. He took part in the botched uprising against British occupation of 1798. The uprising fizzled out before it began. Timmins was arrested, and the court sentenced him to transportation for life as a political prisoner, at the age of forty-three.

He arrived in New South Wales on the *Friendship* in 1800, and received a conditional pardon in 1807 for exemplary behaviour. His pardon described him as: 'Age: 50 years, Height: 5 feet 5 inches, Complexion: dark & sallow; Hair: Black to Grey; Eyes: Hazel'.[9]

In the year of his release, Jack Timmins married a much younger woman, Ann Baldwin, in Sydney. Jack and Ann Timmins would have thirteen children. The most notable was their sixth-born child and third-born son, John (Young Jack), born in 1816.

A census of 1828 reveals that Jack Timmins, aged seventy by that stage, owned 42½ acres at Yarramundi on the Hawkesbury River. First settled in 1794, the Hawkesbury was Sydney's food bowl. Corn, wheat, sheep, cattle and pigs dominated the farming on the river flats.

It is known that Timmins sold eleven bags of wheat grown on his land in 1808. He also bred horses, which were in short supply in the colony in the early nineteenth century; the 1828 census records show that he owned ten of them. But he is best known as a drover. Following the construction of the Great Western Road, he started contract-droving for settlers heading west, then brought stock back east to Sydney as the herds increased.

There has been much speculation that Young Jack joined his father's droving business when a young lad. Research by

historian Bert Howard conclusively disproves that theory. Young Jack did eventually do droving work, bringing cattle from northwest New South Wales to Sydney, but as we shall see, it was as an older man, long after his father had passed away.

* * *

By the 1830s, the sheep and cattle of the fertile Bathurst Plains were flourishing. Bullock drays inched their unwieldy loads of wool bales east over the Blue Mountains to Sydney.

The cursing, bearded men with long greenhide whips who traversed the mountains beside the ponderous bullock-drawn wagons were known as bullockies. It was not the kind of work to imbue a man with a sweet disposition. Bullockies were the most foul-mouthed men on earth – drunken sailors and artillery gunners aside.

Jack Timmins was one of the more prominent droving contractors working out of Bathurst in the early 1830s. He was a man in high demand. No one would have better known the trials of trying to manoeuvre mobs of restive horned cattle over the 103-mile-long, 12-foot-wide gravel road. And Jack Timmins's language on a bad day would have made the bullockies and even the drunken sailors blush.

There was plenty to make an old drover swear on the trip to Sydney. The greatest cause of his profanity was other road users, who were only too capable of spooking obdurate, recalcitrant cattle, which in such confined spaces could have disastrous results.

Jack Timmins soon had the choice of two stock routes over the mountains. The alternative route, the Bell's Line of Road (which opened in 1823), was little more than a track that wound its way from Richmond on the Hawkesbury River to Lithgow on the western side of the Blue Mountains. Because of the problems caused by other road users on the Great Western Road, Timmins and other drovers used the road less travelled when moving stock.

But no matter what route he took, problems were inevitable. Timmins always had one of his fellows riding way ahead to warn oncoming travellers, and to establish suitable passing places.

Losing cattle in the mountains would ruin his reputation, and reaching Sydney with stock losses or poorly conditioned stock would severely diminish his profits. Cattle needed to be quiet and over-conditioned before making the trip, because the mountains were a barren place for decent stock feed. But not every grazier handed fat, quiet stock over to Jack Timmins. The larger the holding, the wilder the cattle. And so, he swore.

The 89 kilometre trip over the range was hazardous, slow going and wearing on man and beast. And the flagging bobtails compounded the difficulties of managing cattle on the narrow road. They were big-bodied, heavy dogs, unable to cope with the heat, or the fourteen-hour work days dealing with stubborn, nasty cattle.

Every serious stockman in New South Wales was over them. And Old Jack Timmins was over them too.

The Timmins name would later become synonymous with a strain of remarkable working cattle heelers, dogs that moved cattle by biting them on their rear hocks (ankles), but that Timmins was Young Jack, Old Jack's son. Young Jack was a dog man through and through, but his father, it appears, was not so gifted. Old Jack struggled with his dogs. Apart from their inability to cope with the physical exertions of the job,

long-coated, heavily built dogs were high maintenance. He was forever cutting sticks and caked muck out of their shaggy coats. Fleas flourished in their thick, matted hair, and he lost dogs to the paralysis ticks east of the ranges every summer. New South Wales was no place for a heavy-coated working dog. Long coats that became matted restricted movement, led to overheating and became an unhealthy burden when wet.

It is known that the bobtails had courage enough, but feet of clay. At every stage camp of an evening, as well as riding watch around the cattle, Jack Timmins would have heated Stockholm tar and dipped the pads of his squirming dogs' feet in it, then pressed their paws into a pan of finely ground cold ashes.

Dog-shoeing was a widespread practice among drovers in Australia. Jack Timmins probably fitted soft kangaroo-hide boots to the dogs with the worst feet every morning, but some dogs hated them and chewed them off. A dog with no feet was worse than useless, and could not impose itself on the cattle, slowing the drover's progress.

In the paddocks of the lush Hawkesbury River flats, the bobtails worked like navvies. But out on the flint-hard Bell's Line of Road, where they needed to attend to their duty from daylight to dark, they failed. Long droving trips ask too much of any dog. Ceaseless work in the harshest environment, bullying horned uncooperative cattle in every kind of condition, is unnatural for a dog. If a difficult mob of cattle walks 10 miles, a working dog runs 30 or more moving them that 10. Jack Timmins's crippled bobtails spent too much time with their worn feet up.

Old Jack Timmins needed better drover's dogs. *Much* better drover's dogs.

Any stockman worth his salt could see there was one dog in New South Wales that had the potential to contribute to a first-class drover's dog. It was medium-sized, courageous and highly intelligent. It had boundless energy, its paw pads were thick and used to harsh terrain, it was strong in the jaw and had a powerful bite, and the heat did not bother it at all.

But, like any dog, it had its drawbacks. First, it was only ever any good as a pup. Second, it was a known sheep-killer. And third, it was actually a wolf.

Old Jack Timmins knew that none of the British working dogs would make the grade as a droving dog in New South Wales. It was a different place altogether, and it needed a different dog altogether.

He was obviously a desperate man. He turned to the enemy for help. In a last-ditch gamble, he gave the dingo a chance to make something of the bobtail.

* * *

A domestic bitch normally comes into oestrus (season, or heat) every six months. She will usually come on heat for the first time at between six and twelve months of age, depending on the individual.

A heat usually lasts about twenty-one days, and consists of three stages, each of around seven days. The first stage sees the bitch develop a swelling of the vulva and a bloody discharge. During this period, she will emit a scent in her urine that is irresistible to male dogs. This scent is carried by the wind and will attract all male dogs that are able to reach her.

Competition for her favours during every stage of the heat is fierce. Males lose all sense of self-preservation, and the most dominant males will fight each other in order to win the right to mate with her. This is especially so if she is out roaming, and many bitches will do their best to escape a yard when in season, such is the strength of the urge to procreate.

Beating up all the competing males will not necessarily guarantee a male the prize of mating with a bitch on heat. Many bitches definitely have their preferred mating partners and some will only 'stand' for a particular male.

Mating will only occur during the second phase of the heat, from about day eight to day fourteen. Leading up to this time, the bitch's discharge will change in colour from blood red to a milky or clear colour. This means the bitch is ready to stand for a dog – which literally means she will stand still and twist her tail to the side and over her back. When she does this, the dog will mount her and they will mate. The male will then lift one front leg and then a hind leg over the bitch's back, so that they are both standing on the ground still locked and joined, but facing away from each other. They can remain locked like this for up to an hour. It gives them protection from other predators and ensures the best possible chance of fertilisation.

Mating can occur several times and with multiple males. It's interesting to note that a bitch with, say, three eggs can mate with three separate dogs: dog one may fertilise one egg, dog two may fertilise a second egg, and dog three may fertilise the final egg. The bitch could then give birth to three puppies, each with a different father. It regularly happens.

The third phase of the heat sees a decline in the bitch's discharge and a decrease in her willingness to stand for a dog. The decrease in discharge attracts fewer dogs, and males begin to lose interest in her.

It is highly improbable that Jack Timmins kept a dingo. In all likelihood, he would have used a wild sire. He would have ridden out with his best bobtail bitch when she was in season and when he thought her 'ripe' and ready to stand. He would have chained her with a tightened collar, probably to a tree on the edge of a creek, fed her up, and left her for a few days. Paw prints within chain-reach would have been proof enough that she had been visited by suitors, and had mated.

As we shall see later, a domestic dog crossed with a dingo will look like a dingo. Red was the predominant, but not only, colour of the dingo, and a bobtail bitch put to a red dingo would probably produce red pups. That first cross would also have sported long tails. But knowing that red brush-tailed dogs would raise concerns with clients and neighbours, Timmins probably decided to dock the puppies' tails, as was the British custom with droving dogs. It is also possible some puppies may have been born tailless. For this reason they became known as red bobtails.

The bobtails matured into dogs that were long rather than tall, short-coated, erect-eared and dingo-red. A tailless dingo. There was no question that they were keen to move cattle. They were full of running, and word spread along the Hawkesbury. There would have been plenty of cattle men who were keen to see how the pups progressed, because anyone who moved stock long distances knew the black bobtail's shortcomings. Interest in the red bobtails was high. Yet Old Jack Timmins had seen enough of both dingoes and bobtails in his time to have reservations about his red bobtails. Would the dingo and the bobtail be the right mix of blood? Only time would tell.

But some of the Hawkesbury locals would not have been happy. Sheep men detested the dingo – and not only the dingo, but any dog that wasn't theirs, and wasn't attached to a thick, short chain. It is easy to imagine them declaring that breeding dingoes with bobtails was a dangerous novelty that would be of no ill consequence to cattle men, but disastrous for the local sheep men.

* * *

Old Jack Timmins was as honest as a day's droving was long. Other than wanting to see Ireland freed of British oppression, he had never committed a crime in his long life. His involvement with the Society of United Irishmen had devastated his middle age. Yet in Sydney he had been a model prisoner and had never felt the lash.

Since receiving his pardon, his fair dealing, his care of his clients' stock and his unfailing reliability had built him an enviable reputation. But his dog-breeding experiment almost destroyed it.

The red bobtails were a disaster. The scanty anecdotal accounts of the time state they were just too savage and attacked calves. It is believed that the red bobtails were destroyed, probably shot on the spot, and the experiment abandoned. When the news spread, the Hawkesbury sheep men would have struck up a chorus of tsk-tsk-ing, and the general opinion that it was culpable lunacy to trust a dingo cross-breed in the first place.

So why was the red bobtail a failure? Theoretically, the bobtail and the dingo could have produced a competent sort of cattle dog, but not in the first cross. Half-dingo was too much dingo. The pups in the first litter would have been very dingo-like in appearance and temperament.

Nothing is known of the training and management of the red bobtails, but they would have needed firm handling from an early age. Were they raised with the naïve expectation that a first cross would just work like a bobtail? More than likely.

The cattle they were trialled on may well have contributed to the problem. If they were worked on wild, flighty stock and the dogs got overexcited, then anything could have happened. It is usual to blame the dog when behaviour goes pear-shaped.

Perhaps the use of a wild dingo sire further contributed to the wildness and lack of tractability. A sire from several generations of captive, tamed dingoes *might* have been a more appropriate choice.

Old Jack Timmins probably saw the type of cattle dog he wanted in his mind's eye, and he was certainly heading in the right direction, but he didn't persevere. Someone else eventually would, and until then, like everyone else, Old Jack Timmins the drover was still stuck with the black bobtail.

On 21 February 1837, Jack Timmins died at the ripe old age of seventy-nine. He never did get to see an improvement in his wretched bobtails.

CHAPTER 8

Hall's Heelers and Timmins Biters Build a Beef Empire

G eorge Hall was a Northumbrian free settler who came to New South Wales intent on becoming a major producer of cattle. He was from a large farming family in Lorbottle, but undertook an apprenticeship as a joiner and carpenter and became experienced in constructing farm machinery. He arrived in the colony with his wife, Mary, and four small children, on the *Coromandel* in 1802. They were part of the group of much-needed tradesmen and farmers who called themselves the Presbyterian Free Settlers.[1]

Governor King came aboard and warmly welcomed them when their ship dropped anchor in Sydney Cove. George Hall, speaking on behalf of the settlers, asked the governor to consider granting all of them land in the same vicinity. King happily agreed, and provided them with temporary plots at the government farm at Toongabbie. A year after arrival, the *Coromandel* settlers were granted fertile land on both banks of the Hawkesbury River downstream from Cattai Creek.

As George Hall improved his 100-acre property, Bungool, he acquired more land. The family moved to Percy Place at Pitt Town and managed their properties from there. Within less than twenty years, George Hall had acquired over 800 acres of productive farmland along the Hawkesbury. His dream of a cattle empire was becoming a reality. George and Mary Hall's family expanded in New South Wales to six sons and three daughters.

The sixth child and fourth son, Thomas Simpson Hall, was born in the Hawkesbury Valley in 1808. He was as driven and ambitious as his father. When he was still in his late teens he ventured north with his older brothers into the upper Hunter Valley, to take up land they had previously scouted for their father. There they established two cattle properties, first Dartbrook, then the much larger Gundebri, close to present-day Aberdeen and Merriwa respectively. Thomas set up his home at Dartbrook and eventually oversaw the entire Hall family cattle empire from there.

George Hall and Old Jack Timmins were Hawkesbury contemporaries. Both came from farming backgrounds. One was an emancipated Irish convict, the other an English free settler; New South Wales was the great social leveller. Both had large families and several sons, and both had a son with an abiding interest in droving dogs. And both sons would make their mark on the development of Australia's working cattle dogs.

* * *

George Hall was always buying, selling and moving stock. Now he needed to buy a lot of cattle to stock his new holdings in the Hunter Valley.

In 1836, a workforce of 170 convict men made that task much easier for him. That year, after a decade's toil, they completed the Great North Road. It started at the Parramatta

Road and wound north to Wiseman's Ferry, on the Hawkesbury River's southern bank. On the northern bank, it ran north through 160 miles of wilderness to the fertile downs of the Hunter Valley.[2]

The isolated Great North Road was unpopular with travellers, but it was the perfect droving route for the Halls' cattle. The Halls travelled north without stock over the Putty Road or Howe's Track, as it was then known, because there were public houses located along its length, and the thirsty drover is not a camel. They brought their cattle back over the Great North Road situated east of Howe's Track. The almost unused Great North Road had little or no public houses so it offered less distractions, and the empty road was far easier for drovers with stock to negotiate than Howe's Track. The new road was a boon to George Hall's cattle business, which was expanding annually under Thomas Hall's astute management. Though a younger son Thomas was given management of the family business from the Hunter Valley northwards because he had the business and stock smarts, and because George Hall had more than enough to keep him busy with his Hawkesbury properties.

The Halls regularly brought their stock south from the Hunter Valley. Initially they walked their cattle to the northern bank of the Hawkesbury and there met the buyers who purchased their stock. The Halls then swam the cattle across the river at Wiseman's Ferry. There on the southern bank they handed over the cattle to the buyers. It is not known why this convoluted form of transaction and delivery was conducted that way.

Selling the stock at Wiseman's Ferry was an unsatisfactory arrangement for the Halls. Their stock had just completed an arduous journey and were short on condition, which resulted in lower prices. In a successful value-adding strategy, in 1835 the family bought land at Liberty Plains (now Auburn, in Sydney's west), and constructed a series of holding paddocks. They were then able to rest and fatten their cattle from the northern properties on their rich Hawkesbury holdings before walking them to their own sale yards. It is interesting that Hall Street, Northumberland Road, Simpson Street (Thomas Hall's middle name) and Dartbrook Road in Auburn mark the apparent boundaries of the Halls' Liberty Plains holding paddocks.

* * *

A surviving portrait of Thomas Hall depicts an intelligent-looking, handsome young man in his mid-thirties, with soft, boyish features. His appearance belies the resolve and determination that were the markers of his indomitable character. Using his father's financial wherewithal, Thomas Hall oversaw an incredible expansion of the family holdings in an astonishingly short period of time. He led expeditions to the north of New South Wales and established holdings that would eventually total over 1 million acres.

Thomas Hall's chain of properties was the first of its kind in New South Wales, and may have been the model that inspired Sir Sidney Kidman to establish the incredible pastoral empire that spanned the continent from Australia's tropical north to Adelaide in the late nineteenth century.

Dissatisfied with the standard of the cattle in New South Wales, Thomas set out to do something about it. He managed the breeding programs for all the family properties from

Dartbrook, developing a polled (hornless) variety of Durham short-horn cattle.

Thomas and his brothers knew only too well the failings of the bobtail as a long-distance drover's dog. Losing 200 head of cattle in rough country between Breeza and Dartbrook may have been the incident that caused Thomas to create a better cattle dog.

Thomas Hall was an experienced stud master; he surely kept stud records for his heelers, like the stud book he kept for his excellent polled short-horn cattle. Yet no record of his heeler breeding program exists.

So, we can only speculate that he collected a pair of unweaned dingo pups – tiny things, a week old – and put them to a nursing brood bitch. The dingo pups would have lived like his other dogs until they neared sexual maturity. Then, to prevent them from returning to the bush, he would have had them kennelled well apart, with domestic dogs of the same sex. He probably chained his dingoes at the stockyards while the bobtails worked cattle. He now had dingoes that were as tame and as 'cattle-educated' as wild things could be.

Finding the right working dog would have been the next challenge. But unlike Old Jack Timmins, Thomas Hall had connections.

Historian Bert Howard was the first person to clarify the heeler myths, identifying a dog he calls the Northumbrian drover's dog or cur as the domestic progenitor of the Hall's heeler.

Sometime in the early 1820s, Thomas and George Hall decided to import Durham cattle from family in Northumbria. George would have written to a relation in Lorbottle with a request to buy family-bred cattle. It is also highly likely that he asked for some of the blue-mottled curs the family had long been breeding.[3]

Robbie Hall, a Hall family historian in Northumbria, has confirmed that the Hall family maintained a line of blue mottled curs.[4] There is little doubt, therefore, that blue-mottled or blue-speckled curs were sent from the Hall holdings in Northumbria with Thomas Hall's Durham cattle.

The cur was common in northern England. Type and breeding would have varied from owner to owner. In *A General History of Quadrupeds*, Thomas Bewick described the cur as a trusty and useful servant to the grazier. He wrote that although it wasn't recognised as a distinct breed, it was the most commonly used type in the north for managing cattle.[5]

The droving cur was a stark departure from the sheep-working shepherd's dog, being developed for much harder work. In describing the cur's behaviour and method of working, Bewick may as well be describing the Australian heelers:

> They bite very keenly; and as they always make their attack at the heels, the cattle have no defence against them: in this way, they are more than a match for a Bull, which they quickly compel to run. Their sagacity is uncommonly great: they know their master's fields, and are singularly attentive to the cattle that are in them. A good dog watches, goes his rounds, and if any strange cattle should happen to appear amongst the herd, although unbidden, he quickly flies at them, and with keen bites obliges them to depart.[6]

The cur in Bewick's engraving has a short coat and a body squarish in profile, with long legs and a lot of daylight under it.

It is bobtailed, has half-pricked ears, a longish neck, a strong, tapering muzzle and typical collie markings. The pictured specimen looks like it means business, and a cur in the middle ground is depicted heeling a bull. It appears to be a fast dog and has large, pronounced feet. It is not hard to image some element of terrier ancestry being present, though it is impossible to identify the now certainly extinct terrier breed or type used back then, it appears that a terrier such as the English white terrier could have been involved. They had a relatively compatible build and would have added determination and muscle.

No depictions of the actual dogs used by Thomas Hall for breeding survive. It's such a pity; much confusion and misinformation might have been avoided. But if posterity was of importance to Thomas Hall, he never made it clear to the executors of his estate.

* * *

So, Thomas Hall created his heeler by crossbreeding the family working cur with the dingo. He would then have selectively bred his dogs until they were displaying all the physical attributes and working traits needed.

As in any breeding program, there would have been unsuitable specimens, and he obviously culled ruthlessly. Yet he achieved his ideal type in quick time. By 1830, Thomas Hall had his heeler, the first successful working dog created in Australia.

Hall's breeding program had produced a medium-sized brush-tailed dog with a rectangular dingo appearance. It was simply the acclimatised dingo, with the cur's working-dog wiring. From the dingo, Hall's heelers inherited cunning, high intelligence, resourcefulness, perfect adaptation to the environment, and a tireless, economical gait.

From the cur, the heeler inherited a powerful work ethic, and the skill of heeling – darting in a crouch behind an uncooperative beast and biting it on the hock of the beast's weight-bearing leg. The heeler also inherited the courage to confront wild cattle front-on, snapping at their heads to turn them around, usually heeling them on their way. But with the boss it was very willing to please. It had the cur's natural suspicion of strangers and the cur's protective devotion to its master, his stock, plant, and property.

Hall's heelers had a protective double coat that was all dingo. It consisted of a fine, dense, insulating undercoat, and a longer, weather-proof outer coat. At first it retained the dingo's red colouring, or a red speckled or mottled variant, with a red undercoat that had solid red patches. But it was the cur's blue colouring that eventually appeared with regular back-crossing. This back-crossing to the cur would have been necessary to strengthen the working instinct in the Hall's heeler. It would eventually produce the occasional tailless dog, but in the early years they were a rarity.

The cur's blue colouring ended up predominating because of Hall's selective breeding, and an obviously strong preference for a non-dingo-coloured dog. His dogs were mottled, speckled or agouti (blue and white hairs evenly intermingled), with or without black patches. They often had tan markings down the legs, on the chest, inside the ears and under the tail, just like the marking patterns found on black and tan dogs.

Colour and markings identify many breeds. Markings can be attractive, but they have no bearing on working ability.

Uniform or attractive markings would have been the last of Hall's considerations.

As Thomas Hall bred his heelers, he would have distributed them throughout his father's Hawkesbury and Hunter Valley holdings and other family properties that stretched northwards through the Liverpool Plains, and the Northern Tablelands region of New South Wales. The Hall pastoral holdings eventually reached as far north as the 305,000 acres held along Balonne River in midwest Queensland. The Halls' station managers assumed responsibility for breeding the heelers needed for their properties.

The Hall's heeler remained the exclusive property of the Hall family, and it was said that no one was going to get their hands on one while Thomas Hall lived.

That policy was good in theory, but it would have been impossible to manage, particularly when the Halls and their employees were on the road droving, or working cattle in Sydney or other towns where free-roaming bitches on heat were common. Any bitch in season would have been a biological magnet for a red-blooded Hall's heeler. Incidental matings happened.

It was that very scenario that produced the temporary type locally called the Smithfield heeler in the 1840s. It was almost certainly a cross between Hall's heelers and other working types, possibly bobtails (confusingly, known also as Smithfields).

Bert Howard's research has identified the existence of this type, but there were obviously not many of them, and while they soon vanished there is still a stubborn tendency for some people to call heelers – particularly stumpy-tailed heelers –Smithfields.

* * *

Sadly, the Hall's heelers on the southernmost properties were only in full-time work for around forty years before transport technology superseded them. After the railway made it to the upper Hunter Valley in 1872, there was less and less call for cattle dogs to perform long-distance droving work. The coming of rail coincided with the introduction of mass-produced wire, followed soon afterwards by barbed wire. Fenced and paddocked properties bred quieter and easier-to-manage stock.

The more remote districts of northwest New South Wales and southwest Queensland became the last bastions of the working heeler.

Hall's heelers were hard dogs, developed for controlling wild cattle in rough country. They not only kept a mob together, but they were also efficient at driving beasts out of thick cover, or recalcitrant steers back to the mob, when they decided to go their own way. This was one tough dog. It took its work seriously – too seriously for quiet stock on smaller, improved holdings. And the increasingly discerning market did not want mauled stock. The Hall's heelers' heyday on the long eastern droving routes had come and gone. And so had the Halls' working dogs' advantage over their competitors.

Thomas Hall died aged sixty-two in 1870, just two years before the railway arrived in the upper Hunter. The Hall's heeler's status as the world's only privately owned working-dog breed ended soon afterwards. Following his passing, the entire Hall rural empire was broken up and sold. Properties, plant, stock, dogs, the lot.

He had been an ingenious, driven innovator, powerful and influential yet modest and unassuming. His had been a life of high achievement, yet he had died with only a modest fortune.

The accumulation of wealth had been unimportant to him. His family was quite wealthy but he chose to live simply and without ostentation. He was a true pioneer, and made an enormous contribution to the fledgling beef-cattle industry in Australia.

Thomas Hall's stud records were consigned to the depths of a disused well when Dartbrook was sold. The exact origins of the early heelers died with their remarkable master.

As innovative and forward-thinking as Thomas Hall was, he could not have imagined what the future held for his heelers. The simple act of preserving his diaries, records and stud books would have clarified the past for future generations, and saved Hall's rightful legacy from self-interest and hijack.

* * *

Initially, beef cattle in Australia had got the jump on wool. Cattle were tough and reasonably self-sufficient. Only the old and weak fell prey to dingoes, so cattle were ideal for large, unimproved holdings. Their biggest drawbacks were their independent nature and their natural unwillingness to cooperate unless bullied into compliance. And when unhappy they could be downright dangerous.

The Halls and their competitors had established a viable beef industry, but events across the Pacific led to wool challenging beef's primacy in Australia.

After the United States Civil War (1861–1865), the battered American nation was desperately short of wool, and it turned to the world's largest wool merchant, Great Britain. Demand in Britain for Australian wool outstripped supply, and high prices provided all the incentive graziers needed. Sheep were a lot more work and required more infrastructure to manage them, but they reproduced much more quickly than cattle and were far more drought-resistant.

Australia had already established itself as the world's premier wool-grower, but with the surge in demand, pastoralists went wool-mad. As wool swept all before it, so the demand for sheepdogs grew and the need for Hall's heelers further declined.

Yet in northwest New South Wales and southwest Queensland, cattle still held sway and it was a region that attracted a great many cattle men. Young Jack Timmins was one of them.

As a young man, (Young) Jack Timmins had found regular employment as a stockman and farmhand with Elizabeth Laycock Matcham Pitt, known as the Widow Pitt, at Kurrajong, a small town in the foothills of the Blue Mountains. Jack struck up a strong friendship with the Widow Pitt's son George, and demonstrated his skill with cattle, horses and dogs. He would remain in the Pitts' employ for several decades.

George Pitt was an ambitious and successful cattle man who set himself up as a small-scale competitor to the Halls. In 1847, he and Jack Timmins, then thirty-one, walked a large mob of cattle from the Hawkesbury to Coorar Station, a 72,000-acre Pitt holding west of today's Moree, which itself lies on part of what was once the Halls' Weebollabolla Station.

The Hall stockmen had been using Thomas Hall's heelers at Weebollabolla and the adjoining Bulleroo Station since 1832, and there is no question that Jack Timmins was aware of them. But it is not thought that he acquired Hall's heelers until, at the age of fifty-five and with George Pitt's assistance,

he took possession of Rocky Holes Station near today's Warialda. This was in 1871, the year after Thomas Hall's death, when Hall's heelers became available to the Halls' former competitors.

Timmins had bred and worked bobtails at Kurrajong and was known to be a great dog man. Like everyone else stuck with the bobtails, he would have been acutely aware of their shortcomings. After the failures of his father's red bobtail, Hall's heelers would have been a revelation to him.

At some stage Timmins must either have bred or acquired one or more tailless heelers. It was from this tailless stock that his reputed line of Hall's heelers, the Timmins biters, was developed.[7]

Timmins biters were like Hall's heelers in every respect. Today, their descendants the stumpy-tail cattle dogs are a slightly more squarish dog like their cur ancestors. Timmins's selective breeding produced a different strain that bred true to type and reliably produced tailless dogs, though tailed dogs are not unknown in tailless lines.

The stronghold of the Timmins biters was the northwest of New South Wales and southwest Queensland – the very country where the Hall's heeler was developed. Conditions there called for a tireless, courageous heeler of semi-wild cattle on large holdings, and for long-distance droving in rough country. Timmins biters became hugely popular with the cattle men and drovers of the region, and would remain so until at least around the 1930s.

The Timmins biters' working style was as unsophisticated and uncompromising as the men who worked them. There was no please and thank you when it came to dealing with obnoxious cattle. This eye-watering description of some

Timmins biters' treatment of an uncooperative beast shows their worth to the cattle man:

> On one occasion at Bogamildi we were yarding cattle from Belalie Camp when a big double X bullock broke away. Jack [Timmins] set his dogs on him. They went straight for his nose. One pinned him by the nose and the other went back between the bullock's legs to nip its testicles. The bullock turned a complete somersault [as you would], got up and shook himself, looked around and galloped back into the mob, bellowing, with the dogs heeling him as he went.[8]

In the northwest of New South Wales and southern Queensland Timmins biters, like their tailed Hall's heeler cousins, enjoyed legendary status for their work ethic, tenacity, intelligence and devotion to master and family. Both varieties also became extremely popular as first-class family dogs and watch dogs with a reputation for being fiercely protective, and for biting first and asking questions later. Bert Howard relates that the cattle families would raise a heeler puppy for their children and that dog's job was to accompany the children while playing and adventuring around their homes perched on the edge of the wilderness. They could hardly have had more dedicated guardians.

The distinctive colouring of these dogs – solid or speckled blue, with black patches and tan markings, or red speckle – became the common theme of the Australian town mongrel as heelers, tailed or stumpy-tailed, crossbred with everything from terriers to spaniels. Heeler blood was valued for hunting dogs, particularly pigging dogs.

By the early twentieth century, wire fencing, rail and road transport was starting to supersede the Timmins biter as a long-range droving and working cattle dog even in the west of New South Wales and Queensland. After Jack Timmins's passing in 1911, his famous dog became popularly known as the stumpy-tail, but its heyday was slowly coming to an end. Over the succeeding decades of the twentieth century it gradually declined in popularity as a working dog.

At the same time, interest in the tailed Hall's heelers was reviving. But it wasn't cattle men, drovers and bush folk who saw their value; there were men in Sydney who couldn't wait to get their hands on them.

These fellows were stock buyers and butchers who wanted harder cattle-working dogs for metropolitan abattoir and saleyard work. The Halls had been walking their cattle into Sydney almost every week in the days before the establishment of the Hunter Valley rail line, and there were also a great many urban enthusiasts who had designs on the breed.

Jack and Harry Bagust and Alexander Davis were the first private breeders in Sydney. They were also involved in the dog-show scene, and as we shall see, their interest in the tailed Hall's heeler would have dramatic and long-term implications for both the tailed and tailless heelers.

* * *

Thomas Hall had died in the same year as the British military garrison finally ceased convict-guarding duty in Australia and left for Britain. Two years prior, the last convicts were transported to Western Australia.

That same year, 1868, a black and tan bitch puppy was born in southern Australia. She was a Scottish working collie, a gathering-sheep worker. For 150 years, her descendants, Australia's collies, would bear her legendary name. They have proven to be the greatest and hardiest working dogs ever created, and it was they who gave Australia the big leg-up onto the sheep's back.

CHAPTER 9

The Making of Australia's Collie

Australia's collie, the kelpie, emerged at a time when Australia's pastoral legends were being written. Wool was selling for record prices, and steamships loaded to capacity with bulging hessian bales sailed almost weekly from Australian ports, bound for the woollen mills of Great Britain.

Everyone was mad about the bush. One of Australia's iconic pastoral images of that era is Tom Roberts's *Shearing the Rams*, painted in 1890, which portrays hand-shearing in a rough-hewn timber shearing shed. But already technology was nipping at the merino's heels. In 1888, around the time Tom Roberts was working on his masterpiece, steam-powered shearing handpieces were being installed on Dunlop Station at Louth, a Darling River town in far western New South Wales. These machine shears, invented by an Irish-Australian, Frederick Wolseley, would revolutionise the shearing process and largely – though never entirely – replace the need for hand shears. Within a matter of decades, every shearing shed in Australia would convert to the new technology.

Yet in the last decade of the nineteenth century, shearing was still done using hand shears in most sheds, and it was grinding, back-breaking work. Shearing legends were made in the days of the hand shears.

In October 1892 at Blackall, a town in western Queensland, a shearer named Jackie Howe wrote himself into Australian folklore. He shore 321 sheep in seven hours and 40 minutes at Alice Downs Station. His tally for the week was 1437 sheep, at an average of 286 a day.

This hand-shearing record has never been bettered. Howe's daily tally was only beaten in 1950 by a shearer named Ted Reick, but he was using machine shears.

Coincidentally, hand shears and Australia's nineteenth-century sheep-working dogs were both superseded at about the same time. The efficiency resulting from both advances would propel wool production in Australia to a truly industrial standard.

* * *

When wire fencing became widely available in the late nineteenth century and stock were able to be contained, the working-dog needs for cattle and sheep were reversed. Cattle confined to paddocks became more accustomed to human interaction and much easier to manage, and as we have seen there was less call for hard-heeling dogs.

Fencing had the opposite effect for sheep. Flocks increased in size to thousands to a paddock, and sheep became more difficult to manage as they became less used to human proximity. Large holdings in the country's interior had paddocks tens and even hundreds of square miles in area. Graziers were mounted, and the sheepdog was required to cover much more ground and work larger mobs of uncooperative sheep.

The inadequacy of once-passable sheepdogs quickly became apparent. Shearing-shed, sheep-wash and yard work processing huge mobs on the big runs required a dog with a powerful work ethic, and matching sheep-smarts and stamina. The long coated Scottish shepherd's dogs, often called working collies, imported into Australia during the nineteenth century were excellent workers, but would have been tested by the rigours of the environmental and climatic extremes of the Australian interior. Burrs, sticks and mud would have inhibited their movement, and rain-soaked and mud-caked coats must have weighed them down.

These working collies were the forebears of the distantly related kelpie and the border collie, but were not identifiable as a defined, standardised breed. They usually sported the half-pricked ears and longish coats common to most Scottish working collies. The common feature was black and tan colouring.

Unlike the development of the Hall's heeler and Timmins biter, there was probably no deliberate attempt to create an Australian version of the Scottish working collie. The consensus seems to be that this breeding process just happened through good fortune and coincidence. It appears that graziers were generally satisfied with their sheepdogs, but naturally, there would have been men striving to improve the working collies to better suit the changing face of Australian wool-growing and the local conditions. Or perhaps there was a contrived effort by some sheep men to pool certain genetic resources and create a superior sheepdog.

Whichever is true, we know that most of the men involved in the early breeding of smooth-coated working collies in Australia knew or had connections with the Rutherford family.

Dingoes are highly adaptable and colonised virtually every region of Australia.
(Courtesy of Northern Territory Library, ABC TV Collection)

Two Pintupi elders with dingo juveniles in the early 1950s.
Neither dingo looks happy at being held. (Courtesy of
Northern Territory Library,
Ted Evans Collection)

Early colonists crossed greyhounds with Scottish deerhounds and called them kangaroo dogs. More powerful than a greyhound and swifter than a deerhound, they had harsh, wiry coats. Their thicker paw pads, which better tolerated the hard, stony ground, stood a better chance of success against dangerous kangaroos. (Courtesy of National Library of Australia, 134298121)

A kangaroo hunt, circa 1901. One bloke is trying to dong the unfortunate marsupial on the scone while another thrill-seeking ghoul is trying to hamstring it. One of the kangaroo dogs seems to have bitten not only the kangaroo, but also the dust. (Courtesy of National Library of Australia, 147416033)

Kangaroo dogs were part of station and rural life. Ladies from St Margaret Island, Victoria, circa 1910, with smooth-coated (left) and rough-coated (right) kangaroo dogs. Both coat types could appear in the same litter, though rough-coated dogs were preferred. (Sydney Pern, Courtesy of Museums Victoria, 110275)

'It is fond of Rabbits and poultry, which it eagerly devours raw; but will not touch dressed meat.' Woodcut of the 'New South Wales Wolf' (the dingo) from Thomas Bewick's *A General History of Quadrupeds* (1790).

An illustration by V Woodthorpe, circa 1802, of the 'Native Dog' (the dingo) of New South Wales. (Courtesy of National Library of Australia, 138471140)

The kangaroo dog became essential to the Indigenous people of Tasmania. This photo from Oyster Cove, Tasmania, shows Trugannini (second from left, seated) of the Nuenonne people of Bruny Island, with her family and a kangaroo dog.
(Courtesy of National Library of Australia, 140385483)

Northern New South Wales grazier Neville Butler with a stumpy-tail Hall's heeler around the turn of the nineteenth century. This dog may possibly have been a Timmins biter, a popular local strain. (Courtesy of Bert Howard)

The Allan family, pictured here with their Hall's heeler circa 1900, were from Aberdeen in the Hunter Valley, New South Wales. It is thought the Allans were Hall's employees. (Courtesy of Bert Howard)

This is the earliest known photo of a Hall's heeler, circa 1880. The image of the dog has been very badly cropped and modified some time ago. It is not a flattering portrait of the breed. (Courtesy of Bert Howard)

King's Kelpie, the daughter of Gleeson's Kelpie, and the dog considered by many to be the matriarch of the kelpie breed. She tied with Gibson's Tweed to win the historic Forbes sheepdog trial of 1879. (Courtesy of Tony Parsons)

Coil, the great dog who started the Australian working dog narrative. Gently coaxing a little chicken into a jam tin (out of view) was nothing compared to winning the prestigious Sydney sheepdog trial in 1898 with a perfect score — on three legs. (Courtesy of Tony Parsons)

Sydney-based journalist Robert Kaleski, Australia's prominent dog authority, with one of his Australian cattle dogs.

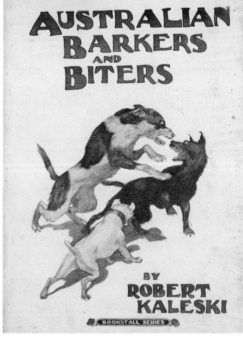

First edition cover of Kaleski's *Australian Barkers and Biters* (1914), responsible for propagating myths and misinformation about Australia's working dogs and their origins. (Courtesy of State Library of New South Wales, Mitchell Collection)

This iconic photograph graphically demonstrates why cattle dogs, stumpies in this case, are too severe on quiet stock.

A modern stumpy-tail cattle dog, a derivative of the Timmins biter, which was a tailless version of the Hall's heeler founded by Thomas Simpson Hall around 1830. (Copyright Wendy Hodges, licensed under CC Attribution 2.0 Generic)

Athletic and intelligent, the modern kelpie is the ideal working stock dog. Pictured is the late Karmala Akubra. (Courtesy of Jan Lowing)

This award-winning photo of Akubra demonstrates the focus and devotion to duty that has made the kelpie famous around the world. (Courtesy of Jan Lowing)

Akubra is the model for the bronze statue 'Kelpie' by Bodo Muche, installed at the Australian Stockman's Hall of Fame in October 2013 to 'work' the bronze Merino ewe and lamb statues installed there 25 years before. Both Akubra and Muche passed away within a week of each other in December 2017. (Courtesy of Jan Lowing)

Harada and Tess of the New South Wales Police Dog Unit, who served with distinction despite Alsatians being banned from importation into Australia from 1929 to 1974. (Courtesy of National Library of Australia, 159856193)

Kaspar and Tess leap through a fiery hoop. There was almost nothing Scotty Denholm and the other members of the Police Dog Unit could not teach their charges. (*The Cold Nose of the Law*, C Bede Maxwell, 1948)

Zoe the white Alsatian passes her driving test at the Domain in Sydney and is issued with her own licence. She drove her car via remote command. (*The Cold Nose of the Law*, C Bede Maxwell, 1948)

* * *

As we saw earlier, the best working shepherd's dogs came from Scotland, and varied according to the districts in which they were developed. The late eighteenth and early nineteenth centuries were a period of great change in Scotland's Highlands. Small crofts were turned into large sheep farms and the people moved on. Many emigrated to Canada, the United States, New Zealand, and Australia. English and Scottish shepherds from northern England were recruited to the Highlands to manage the cheviots, and with them came their shepherd's dogs which were ironically of Scottish origin.

Gideon Rutherford was one such man. He and his son, Richard, became prominent working-collie breeders in Sutherlandshire in the mid-nineteenth century, and became known in Australia for their excellent sheepdog bloodlines. Should any one man be accorded the status of the father of the Australian collie that became the famed kelpie, it is Richard Rutherford, who developed a strain of smooth-coated collies for his connections in the Australian wool industry.

His brother John emigrated to Australia in 1848, and in 1849 he began working on Golf Hill Station near Geelong in Victoria under a contract with the Clyde Company of Glasgow.[1] He subsequently became a highly successful businessman and wool-grower. He would be hugely influential and a central figure in the early stages of the development of the kelpie through his brother Richard. Seeing the shortcomings of long-coated working sheepdogs in Australia, he may have written to his dog-breeder brother and enquired about the possibility of developing a smooth-coated collie suited to Australian conditions.

The kelpie did not emerge as a distinct breed until sixty years after the development of the Hall's heeler. Both were New South Wales natives, but for the average Australian, the kelpie was much easier to warm to than the serious, suspicious, hard-biting heeler.

Some very little-known and remarkable coincidences lie behind the creation of the kelpie and the Hall's heeler. George Hall, the father of Thomas Hall; and Gideon Rutherford, the father of Richard Rutherford, were born and raised within about 30 kilometres of each other as the crow flies. The English–Scottish border, the River Tweed, separated their respective homes. Both married local women, both had a first-born child named Elizabeth, and both had sons who created working dogs that would change the course of Australian pastoralism.

Historian Bert Howard and Australia's foremost authority on the kelpie, Tony Parsons of Karrawarra Kelpie Stud fame, have delved deep into the breed's history, and that information is contained in Mr Parsons's definitive work, *The Kelpie*.[2] Unless otherwise noted, all the following information on the kelpie's development is drawn, with Mr Parsons's kind permission, from his impressive work.

While much is known about the kelpie's origins, the truth of its development has been muddied by contradictory theories and origin claims ranging from probable to doubtful to ridiculous. Despite exhaustive research, one major question remains unanswered. The truth is that the mystery may never be conclusively solved. There are however, a couple of plausible theories worthy of consideration. We'll look at them in the next chapter.

Here then, is a summary of the development of the dog that became the kelpie up to the year 1880, as provided by Bert Howard in Tony Parsons's *The Kelpie*.[3] It is a twisting, turning story with a lot of players, though for ease of reading I have tried to simplify it as much as possible.

* * *

George Robertson (1803–1890) was a Scottish immigrant and a successful grazier. He established a large sheep property, Warrock Station, near Casterton, in southwest Victoria.

George Robertson and his wife married late in life and remained childless. With no direct heir to the considerable estate they had created in Australia, they invited George's nephew, George Robertson Patterson, to live at Warrock and assist with the management of the property. From that point the Robertsons treated him as their own child. He was fifteen when he started life at Warrock.

In 1867, George Robertson produced a litter of collie puppies from a pair of imported Scottish working collies of the Rutherford strain. George Robertson and George Patterson agreed that none of the bitch puppies in the litter would be sold until their breeding program was fully established. Interestingly, this was George Robertson's first foray into keeping working dogs. Warrock had traditionally used shepherds to manage its flocks.

At the same time, Jack Gleeson, a young itinerant stockman of Irish descent, was working on the neighbouring Dunrobin Station. Jack Gleeson knew George Patterson and his uncle, and being a dog man in need of good sheepdogs, he took an interest in the Warrock litter.

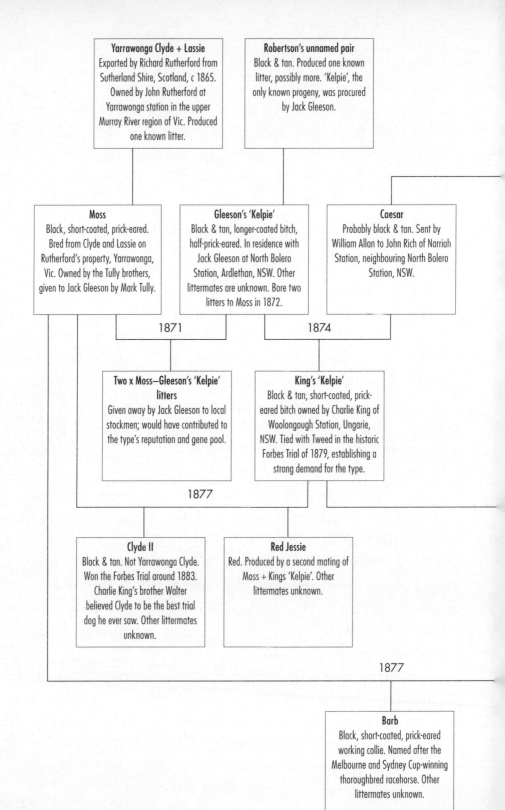

Yarrawonga Clyde + Lassie
Exported by Richard Rutherford from Sutherland Shire, Scotland, c 1865. Owned by John Rutherford at Yarrawonga station in the upper Murray River region of Vic. Produced one known litter.

Robertson's unnamed pair
Black & tan. Produced one known litter, possibly more. 'Kelpie', the only known progeny, was procured by Jack Gleeson.

Moss
Black, short-coated, prick-eared. Bred from Clyde and Lassie on Rutherford's property, Yarrawonga, Vic. Owned by the Tully brothers, given to Jack Gleeson by Mark Tully.

Gleeson's 'Kelpie'
Black & tan, longer-coated bitch, half-prick-eared. In residence with Jack Gleeson at North Bolero Station, Ardlethan, NSW. Other littermates are unknown. Bore two litters to Moss in 1872.

Caesar
Probably black & tan. Sent by William Allan to John Rich of Narriah Station, neighbouring North Bolero Station, NSW.

1871

1874

Two x Moss–Gleeson's 'Kelpie' litters
Given away by Jack Gleeson to local stockmen; would have contributed to the type's reputation and gene pool.

King's 'Kelpie'
Black & tan, short-coated, prick-eared bitch owned by Charlie King of Woolongough Station, Ungarie, NSW. Tied with Tweed in the historic Forbes Trial of 1879, establishing a strong demand for the type.

1877

Clyde II
Black & tan. Not Yarrawonga Clyde. Won the Forbes Trial around 1883. Charlie King's brother Walter believed Clyde to be the best trial dog he ever saw. Other littermates unknown.

Red Jessie
Red. Produced by a second mating of Moss + Kings 'Kelpie'. Other littermates unknown.

1877

Barb
Black, short-coated, prick-eared working collie. Named after the Melbourne and Sydney Cup-winning thoroughbred racehorse. Other littermates unknown.

Brutus
Black & tan, smooth-coated, prick-eared. He won trials at Young, NSW, in 1871 & 1872. After the death of Gilbert Elliot in 1874, William Allan moved to Queensland taking Brutus with him.

Jenny
Black & tan, long-coated, half prick-eared. Nothing more is known of her or the progeny of future matings.

1870

Laddie
Assumed to be black & tan, short-coated, prick-eared. Second infusion into the kelpie strain.

Nero
Probably black or black & tan. Remained in NSW Riverina region.

Mylecharane's dog
Tan. Phil Mylecharane recorded nothing about his dog other than what is here; however, the aim always was to produce short-coated and prick-eared dogs. A 'splendid worker', stolen at Goolagong near Forbes, NSW.

1877

Sally
Black & tan, short-coated, half-prick-eared.

The kelpie family tree.

He tried several times to purchase one of the puppies, a bitch named Kelpie, after a mythical Gaelic water spirit. George Robertson repeatedly refused him, but his nephew was taken by a striking stock horse Gleeson owned.

The Irishman, not one to take no for an answer, saw his opportunity. A surreptitious deal was struck between him and Patterson. One night in early 1869, against his uncle's express wishes, George Patterson swapped Kelpie for Jack Gleeson's stock horse on the bank of the Glenelg River, opposite Warrock Station.

Jack Gleeson wasted no time in making off with his new prize. George Robertson was a powerful man who would not have taken the loss of Kelpie with good humour. So, leaving George Patterson to explain the absence of Kelpie and the sudden appearance of his impressive new mount, Gleeson and Kelpie cut their own tracks north towards Edenhope. They then travelled through the Little Desert region of western Victoria, eventually finding work on Launcelot Ryan's Ballerook Station, 20 kilometres northwest of Kaniva.

It was during the shearing at Ballerook that year that Jack Gleeson broke Kelpie in to sheep work, and she caught the attention of another Ballerook stockman, Mark Tully. It's a small world. Mark Tully was from Sutherlandshire in Scotland, and the Tullys were neighbours of the Rutherfords. Two of Mark's brothers, Walter and Robert, worked on Illillawa Station in southwest New South Wales, owned by John Rutherford and Company. There was obviously a close relationship between both families.

The Tullys were talented stock and dog men. Their own line of highly regarded working collies, known locally as Tullies, had been drawn from Rutherford dogs. The Tullys' contribution to the development of kelpies in Australia has been undervalued. Their dogs were much admired for their working ability and, being of the same stock and type, must have made a huge contribution to the 'kelpie-type gene pool'. The kelpie could just as easily have become known as the Tully.

True dog men and women, no matter what breed they are involved with, know potential and good breeding when they see it. The meeting of Jack Gleeson and Mark Tully might have been coincidental, but their common interest in working collies would be the catalyst for the development of perhaps the greatest working dog the world will ever see.

* * *

At the same time as Kelpie caught Mark Tully's eye, one of Launcelot Ryan's daughters, Mary, caught Jack Gleeson's eye, and he hers. Jack and Mary began to court, but the budding romance was put on hold in early 1871, when Launcelot Ryan and family relocated to Wallandool Station in the Riverina region of New South Wales.

Mark Tully also departed Ballerook and took up a position as overseer on Illillawa, reuniting with his brothers. Jack Gleeson decided to follow Mary Ryan, and did so via Illillawa, taking Kelpie with him.

Mark Tully, with whom he had struck up a friendship, may have invited him, and the visit was probably dog-related. Mark Tully knew his dogs, and he may have told Jack Gleeson about a dog his brother had that he thought would suit Kelpie as a sire. This dog was Moss, a black, smooth-coated, prick-eared Rutherford collie, a Tully.

Jack Gleeson did not stay long on Illillawa. He soon left to pursue his romance with Mary Ryan, but stopped at Goonambil Station, another Rutherford property, along the way. From Goonambil he travelled to Walbunderie Station, a property adjoining Wallandool, where he found work as a stockman. Mary Ryan was then his next-door neighbour, and they picked up where they had left off the year before.

Mark Tully took a job at Goonambil soon afterwards, and Jack Gleeson visited him there in late 1871. When he left Goonambil after the visit, Jack Gleeson had Moss trotting beside his horse all the way back to Walbunderie. Mark Tully had given him the collie sire he needed.

Always on the lookout for better opportunities, Jack Gleeson took a job as the overseer of North Bolero Station, northwest of the Riverina town of Ardlethan. The owners were Patrick and Mary Quinn, immigrants from County Tipperary, Ireland, where Jack Gleeson's family also hailed from. Old country ties ran deep in rural Australia in the nineteenth century. So did kelpie ties. As we've seen, the Quinns' son John (Jack) and his kelpie Coil would later become household names.

Kelpie had two litters to Moss at North Bolero in 1872, and Jack Gleeson gave every puppy away. They'd have all been claimed by local stockmen before they were even born.

* * *

At this point the story shifts to another part of the New South Wales Riverina. Around 1870, friends Gilbert Elliot and William Allan bought Geraldra Station, near Cootamundra. They imported a pair of prick-eared, smooth-coated

Rutherford collies from Scotland, Brutus and Jenny. The pair mated on the journey to Australia and a litter was whelped soon after arrival at Geraldra. Three dog pups from the litter were Laddie, Nero and Caesar.

In February 1874, Gilbert Elliot was killed in a buggy accident. His widow and children returned to England soon after his death, and William Allan sold Geraldra and took up property in the Culgoa district of southwest Queensland.

Allan took Brutus with him, and the rest of his dogs he passed on to friends in the district. Laddie went to John King and his nephew Charlie King, who later settled on Woolongough Station near Ungarie, New South Wales. Caesar went to John Rich, who owned Narriah Station, next to North Bolero Station.

At North Bolero in early 1875, Jack Gleeson's Kelpie whelped a litter sired by Caesar. A bitch puppy from the litter, named Kelpie after her mother, went to Charlie King at Woolongough Station. To avoid confusion, they called her King's Kelpie. Her mother was called Gleeson's Kelpie, then Old Kelpie.

In late 1876, Charlie King put King's Kelpie to Laddie, Caesar's brother, and that union produced a bitch named Sally. Sally was mated to Jack Gleeson's Moss, and that mating produced the famous black dog Barb.

By that time, Jack Gleeson had become the manager of the neighbouring Yalgogrin Station, and finally married Mary Ryan at her father's property, Wallandool. The Gleesons moved to Lake Cowal Station, but soon afterwards, Jack Gleeson contracted hepatitis. Seriously ill, he moved with Mary to Warrnambool in western Victoria, to be near his family.

Before leaving the Riverina, Jack gave Old Kelpie to a friend, Mr T J Garry of Ungarie Station, but she eventually developed cancer and was put to sleep at Woolongough Station. Jack sent Tully's Moss to Charlie King on Woolongough, but sometime later the famous old black collie was found dead on his chain one morning, apparently from natural causes. Meanwhile, Laddie should have kept his mind on the job and left the Woolongough kangaroo dogs to do their work; he was drowned by a kangaroo in a waterhole.

Sadly, Jack Gleeson died on 29 August 1880, just three months after Mary bore him a son, and only ten years after he gained possession of Kelpie from George Patterson. Gleeson had never been in it for the money. He gave away every dog he bred.

In the historical context Jack Gleeson is not considered the originator of the kelpie breed, but he was certainly the catalyst that allowed a group of Riverina stockmen to selectively create a breed from a group of top-quality sheepdogs similar in type.

The kelpie came into being because a fortunate set of circumstances brought several smooth-coated, prick-eared, Rutherford-strain, Scottish sheepdogs and a handful of great Australian dog men together. The right dogs. The right men. The right place. The right time. It has never happened before, and it will never happen again.

Between them they consolidated a gene pool of superior sheep-working stock. A couple more decades of smart selective breeding, hard culling, natural selection and physical adaptation saw the evolution of a true Australian collie that could more than cope with the challenges of huge workloads and the cruel Australian interior.

It was only fitting that the eventual breed was named after its matriarch, King's Kelpie, in turn named after her mother,

Gleeson's Kelpie. King's Kelpie's offspring were in high demand, and were originally known as Kelpie's Pups. Before long, other working collies of similar breeding became known as 'kelpies'. Their reputation as workers spread not only throughout sheep country, but all over Australia.

It is impossible now to trace the breeding programs and hand-to-hand, station-to-station distribution of the early kelpies throughout New South Wales and the other States of Australia. But such was the appreciation of their working ability that it is known to have happened quickly.

Pound for pound, breed for breed, the working kelpie is the hardiest dog in the world. For sheer endurance, courage and stick-with-it-ness in the most trying and adverse conditions on earth, no other breed comes close.

* * *

Right from the breed's inception, famous kelpie progenitors were being entered into, and winning, sheepdog trials. They attracted national attention in both the bush and the cities.

These trials have been a part of the Australian pastoral scene for almost 150 years. The Burrangong Pastoral and Agricultural Show hosted the first trials at Young in New South Wales on 17 April 1871. That was over two years before the first recorded trials in Britain, which took place at Bala in North Wales in October 1873.[4]

Trials were a brilliant concept, and the ideal vehicle for breeders of sheepdogs to test their dogs in open competitions designed to reflect actual working conditions. Trials vary in format, but in Australia the three-sheep trial has traditionally been the most popular. The aim is for the dog to work under

the handler's control to move three sheep through a series of obstacles set out in an approved course, culminating with the sheep penned in a small yard.

There have always been sheepdog trial detractors. Right from the start, some stockmen claimed that a competent trialling dog did not necessarily make a good working dog. In some cases that might have been so, but legendary kelpies such as Coil, Wallace, Biddy, Biddy's Daughter and Biddy Blue were trial champions *and* great working dogs. Their offspring worked some of the biggest sheep stations in Australia. Jack Quinn once commented that his dogs 'weren't kept in lavender for trials. They did ordinary station work, mustering, and so on, just the same as any other station dogs.'[5]

The *Australian Town and Country Journal* of 29 April 1871, reported on the first trial staged by the Burrangong Pastoral and Agricultural Association:

> In sheepdogs there were three entries, and the prize taken
> by a dog the property of Elliott and Allan. The
> performance of this dog was something wonderful. Three
> sheep were let loose and taken outside the ground and the
> dog, upon the word being given, brought them into the
> ground and across through the crowd of people, running
> here, there, and everywhere in a manner which would
> confuse a human being, to their pen, without even so
> much as a bark.[6]

That dog was Brutus. Elliot and Allan, W Rutherford and the Wallis Brothers entered three dogs the following year. The *Burrangong Argus* of July 1872 reported:

The trial between sheepdogs, two of which are Mr Allen's and Mr W Rutherford's entered the lists – was very interesting. The animals were both thoroughly under control and evidently masters of their business, and through a long and severe trial many were the bets hazarded as to how the judges Messrs Todd and Hammond, would give their decision. The preference was, however, in the end given to Mr Allen's entry the same dog (Brutus) which took the prize and even frightened all competitors from the field last year.[7]

In 1873, Gilbert Elliot's 'Scotch collie dogs' Caesar and Jessie received an honourable mention.

It is arguable that the kelpie story really began in 1879, when King's Kelpie, at the age of four, competed in a sheepdog trial at Forbes. The weather was atrocious; it poured rain all day long. The wool-growers of New South Wales and further afield stood up and really began to take notice after King's Kelpie and Gibson's Tweed tied to win the trial. The following account appeared in the August 1879 edition of the *Australian Town and Country Journal*:

> The pastoral and Agricultural show has been fairly successful, notwithstanding the unfavourable weather ... At the trial of sheep-dogs today, there were seven entries, including some of the best dogs in the colonies. After some severe tests the judges divided between Mr Charles King's Kelpie and Mr C F Gibson's Tweed ... Flockmasters came from distances of 150 miles to see the trail [sic], and avowed that it was the grandest contest they ever saw. The dogs worked one and three sheep respectively, and notwithstanding the continuous

rain, some hundreds of people watched the trials for six hours with unflagging interest.[8]

* * *

There have been many great kelpies and many great kelpie men, but no partnership has ever exceeded the fame of Jack Quinn and his great blue kelpie Coil.

Jack was the son of Patrick and Mary Quinn, the owners of North Bolero Station. It was there that Jack Gleeson worked as an overseer, and that Kelpie had two litters to Moss, the Rutherford dog Gleeson got from Mark Tully. Jack Quinn got the best of starts with his working collies, but he also had the wherewithal to do them justice.

The family took up Springfield Station near Cootamundra, where Jack Quinn would remain for most of his life. His obituary would note that 'In his day Jack Quinn was one of the vigorous types of bushmen who loved to put on the boxing gloves, play cricket, run, and jump; and particularly was he fond of a good horse and sulky in the pre-motor days.'[9] But quality working kelpies, bred from the original Kelpie, were his stock in trade.

Quinn paid the hefty price of £20 for a black and tan dog named Clyde which he bought from the Willis family of Junee. He won the Temora sheepdog trials with Clyde in 1890. He then put Clyde to his bitch named Gay, and that mating produced Coil.[10]

Constant sheep work on Springfield Station honed Jack Quinn's kelpies for trials. Genes are almost everything, but as we've seen, a dog is what you make it, and Jack Quinn made a lot of very famous and influential dogs.

Quinn's dogs were soon winning country trials and he was becoming known as one of the best kelpie men in New South Wales. The first Sydney Trial was held in 1896. He entered Gay and she won the inaugural event. The following year he placed second with another bitch, Rose. Then in 1898 he brought his blue dog Coil to Sydney, where Coil performed his legendary two perfect runs, the second with a broken and splinted foreleg.

The new nation had a national hero in the little battler from Cootamundra who had won the Sydney Trial on three legs. The Immortal Coil's victory was symbolic of the crippled wool industry's courage in the face of adversity. It was one of the few bright spots in the bleak years of the devastating Federation Drought.

The kelpie is certainly a hardy dog. It's incredible to consider that it went through much of its development, in fact what many consider to be its golden age, while eastern Australia was labouring under the Federation Drought. That speaks volumes for the kelpie's ability to cope with trying conditions. But drought might have been the furnace the kelpie needed to harden it to the Australian environment.

How many great working kelpies were lost during that environmental disaster we'll never know, yet the Australian wool industry, as good a doer as the hardy merino, would somehow survive. Things were about as grim as they could get, but the market remained as strong as ever and wool bales were still selling like hotcakes.

The *Sydney Mail* of 16 October 1897 enthused:

> One of the most remarkable sights in Sydney is that of the
> Wool Exchange in full blast. Both buyers and brokers are

men of weight and substance and responsibility in the
community, but when the auctioneer puts up a lot the
buyers spring to their feet, wave their catalogues over their
heads, and shout their bids frantically at the seller. It is not
unusual for 100,000 lb worth of wool to be put through
during an afternoon.

Australia's economy at the turn of the twentieth century was
almost wholly agricultural. There weren't many Australians
who weren't directly or indirectly dependent on the fortunes
of wool, mutton, beef, wheat, sugar or the other great primary
industry, mining.

Australians had developed a reverent respect for the merino,
and why not? It fuelled the economy and brought Australia
prosperity and world standing. But it was difficult to truly love
the sheep, except perhaps in terms of a Sunday roast leg. Was it
hard to love the merino because the merino did not love
people? Or due to the common perception that the merino is
not the sharpest tool in the shed? Surprisingly, although the
nation rode to prosperity on the merino's back, it was not the
nation's favourite animal.

Nor was Australia's darling either of our national emblems,
the kangaroo and the emu. They were considered to be
nothing but pests, and competitors of the worthy but
intellectually challenged merino. Australia's favourite animal
wasn't even the cuddly koala, nearly shot into extinction for its
luxuriant fur.

No, the most popular animal in Australia, after the
budgerigar, was the kelpie.

The kelpie is the perfect blend of every working-collie
strong point and native adaptability. Partnered with some of

Shanahan's Loo. Arguably the greatest photo ever taken of a working kelpie. She was a hugely influential bitch and many great modern dogs trace to her. (Courtesy of Tony Parsons)

the greatest ever dog men, the development of the kelpie represented the pinnacle of working–dog design and all–round ability.

* * *

While the kelpie was being developed in Australia, the dog that became recognised as the border collie was being refined in Scotland. It was an amalgam of several different collie types found around southern Scotland and northern England, and was officially recognised as a set type and breed in 1906. It is a remarkable worker, and has won admirers all around the world for its intelligence, obedience and ability to perform almost any task.

The border collie's introduction to Australia had an appreciable effect on the kelpie, and set in train debate and comparison that continues today.

It began when the prominent kelpie stud partnership of King and McLeod imported a very successful border collie dog named Tweed of Roxburgh just after the turn of the twentieth century. There were claims that the infusion of border collie into the kelpie improved the kelpie's working ability. It was a prime example of trying to fix a working dog that wasn't broken, but different stockmen had differing opinions.

Messing around with cross-breeding to fix a localised 'problem' usually has an impact on a large part of the breed population for a long, long time. Yet cross-breeding has had the reverse effect too. Some Australian working border collies, particularly the smooth-coated strains, carry kelpie blood. They are a lighter dog than typical border collies and they have greater ability to endure Australian conditions.[11] While both breed camps have various technical concerns about the influence of the other blood on their breed, the addition of other genes may not be too harmful for either. The kelpie and the border collie share the prize as the best sheepdogs in the world.

The clearest differences between the kelpie and the border collie are temperament and attitude to work. Tony Parsons believes the kelpie generally possesses a high degree of independent problem-solving ability, and wants to make its own decisions. Whether in the paddock or on the trial ground, the kelpie thinks it knows better than its boss. That is a fairly Australian trait, admittedly. In the same vein, the kelpie has a 'she'll be right' attitude, not minding if it makes a mistake so long as it gets the job done. It is a terrific all-round worker but lacks the willingness to please of many good border collies.

A South African kelpie man put it to Tony Parsons like this: 'If you give a border collie a command, he will look at you and ask, "What next?" If you give a kelpie a command, he will look at you and ask, "What for?"'[12]

The border collie is drawn from earlier working collie breeds used to smaller, softer holdings. It is a very focused, dependent breed that has always been more regularly handled and worked. The kelpie is more a creature of the wide-open spaces, developed to manage mobs of sheep on holdings a quarter the size of Scotland. Thinking for itself has always been part of that job.

Border collies do not have the kelpies' endurance, or their ability to take the punishment of working massive mobs in the harsh Australian environment. They are so naturally obedient and trainer-focused that they are better suited to high-demand activities. They make sensational three-sheep trial dogs, and

based purely on the results of the national trials over the last fifty years, they are better than the kelpies put up against them. Border collies also make very good general-purpose sheepdogs. The New Zealanders are so happy with them they won't even look at a kelpie. But for all-round Australian station work, the border collie cannot hold a candle to the kelpie.

* * *

The kelpie was the sheepdog that just kept on giving, and it won the hearts and minds of stockmen who were not easily impressed. These live-wire workers toiled on remote stations in tireless anonymity, earning a reputation for their biddability, intelligence and durability. Yet what was one of the kelpie's greatest attributes, its ability to quickly adapt to its environment, was its most underrated. Any tendency to rose ears, thin paw pads, grossness or a long coat quickly gave way to the forces of selective breeding and natural selection.

The working kelpie's versatility and adaptability are now increasingly recognised worldwide. The hard work and goodwill of Tony Parsons, Tim Austin of Elfinvale Kelpies and many other prominent kelpie stud masters have seen close ties formed with committed overseas supporters. There are now great kelpie men and women, and great kelpies, all around the world.

Today, kelpies work sheep in California, cattle in Arizona and Texas, sheep and Boer goats in South Africa, and reindeer in Scandinavia. One can't help feeling a great sense of pride at seeing the mighty little kelpie winning the hearts of stockmen and women right around the globe.

Sheep, cattle, goats, poultry, reindeer; Australia, the United States, Scandinavia, Western Europe, South Africa; it doesn't matter what, and it doesn't matter where. All that matters to the kelpie is that it has a job. Because it's the kind of dog that believes a job worth doing is a job worth doing properly. Even on three legs.

CHAPTER 10

The Missing Piece of the Kelpie Puzzle

Most of the working-dog breeds we know today are relatively recent developments. The dog has an extremely elastic genetic make-up, and has constantly evolved through breeding or natural selection to suit man's various needs. In one working lifetime, a determined selective breeding program can make significant changes to a type or strain.

Hunting dogs developed for specific methods had no real need to evolve, so they have remained essentially unchanged. Conversely, the working dog, because of the changing nature of agriculture and stock work, was subject to continual tinkering. This resulted in the development of micro-types, small populations of working dogs bred to suit the needs of a family, or local region.

The Industrial Revolution and the expansion of the British Empire began to change the British working dog, particularly in Scotland. As small family tenancies disappeared, stock-working micro-types either disappeared or merged into widely distributed macro-types, which themselves gave way to a handful of major breeds by the start of the twentieth century.

When trying to decipher modern working-dog breed histories, we are handicapped both by a lack of record-keeping and by visual evidence that can easily be misleading. We further hobble ourselves by thinking of British working dogs from the past in terms of the popular breeds we are familiar with today.

Things weren't the same 150 years ago. The few surviving British working-dog breeds are fusions of many different micro-types that will never be known to us. Even over the last 100 years, several established working breeds in Britain have disappeared. It seems the older the working breed, the more specialised it was, and hence the more vulnerable. Since the early twentieth century the remarkable border collie has swept all before it, making appreciable inroads in Australia and even influencing the bloodlines of many working kelpie strains.

It is now known that the progenitors of the kelpie were Scottish working collies, with the main influence being prick-eared, smooth-coated dogs of the Rutherford strain. What is still open to speculation is what smooth-coated working sheepdog type Richard Rutherford used to produce his smooth-coated, prick-eared collies – such a divergence from the typical long-haired, half-pricked or rose-eared Scottish working collie.

* * *

One theory difficult to dismiss was put forward by Tony Parsons in *The Kelpie*. It involves an unexpected source for the kelpie's 'missing link': the German shepherd dog.

Prior to 1899, when a speciality German shepherd club was formed, the working sheepdogs of Germany were a landrace – localised, non-standard – breed that were smaller and more collie-like than today's German shepherd. They were smooth coated, and some bore a truly startling resemblance to the kelpie. Their common coat colouring was black and tan or black, and saddle-marked (marked over the centre of the back) or sable (black-tipped): colouring also seen in the kelpie. It is known without doubt that Gleeson's Kelpie was saddle-marked, and that colouration still appears from time to time in modern kelpies.[1]

It is entirely possible that working German sheepdogs made their way to Scotland with several flocks of Saxony merinos that were imported to Scotland from Germany in the early nineteenth century.

One settler who brought such sheep to Australia was Eliza Forlonge, a pioneering pastoralist born in Glasgow, Scotland, in 1784. Her husband, John, was a merchant, and although she had at least six children, by the mid-1820s disease left her with only two sons, William and Andrew.

The Forlonges decided to become wool-growers and emigrate, as so many Scottish sheep farmers did, to Australia. Wool from merino sheep in the German kingdom of Saxony was realising high prices, so, leaving her husband at home to take care of business, Eliza went to Leipzig with her sons to study sheep-rearing and wool-classing.

Between 1828 and 1830, Eliza and her sons travelled throughout Saxony buying sheep. They purchased 100 Saxony merinos, walked them to Hamburg and had them shipped to Hull in England. From Hull, Eliza and her two sons walked them to Scotland, where the Australian Agricultural Company bought the lot.

Eliza and her sons twice repeated the Saxony sheep-buying journey. She selected the sheep that travelled to Australia with her son William in the *Clansman* in 1829, as well as those accompanying the rest of the Forlonge family in the *Czar* in 1830.[2]

It is highly likely that Eliza and her sons collected sheepdogs along with the sheep they purchased in Germany. An essential part of learning sheep husbandry would have been training and working dogs, and local sheepdogs used to herding merinos would have been an indispensable aid in Eliza's remarkable endeavours.

The story of Eliza Forlonge does not prove Tony Parsons's theory that the German shepherd was the Rutherford's mystery out-cross (unrelated breed). But considering the appearance of the German sheepdog at the time, the possibility that such dogs made their way to Scotland with flocks of merinos certainly lends that theory considerable weight.

* * *

Another more widely sown but less plausible theory is that described by the prominent, but at times fantasy-prone, British border-collie and working-dog author Iris Combe in her popular book *Herding Dogs*.[3] While it appears that much of Mrs Combe's breed-origin theories are apocryphal, she nevertheless sweeps us along to the Isle of Skye, in the Scottish Hebrides when she holidayed there with her husband after the Second World War. There she says they met a retired cattle farmer and drover whom she names only as 'John'. He invited the Combes to his remote croft and they talked about the working dogs he used while still an active farmer.

Mrs Combe says John called his dogs 'kelpies', not collies. He said his kelpies had been as tall as his knee, and lightly built with bear-like hair. Their coats had been harsh to the touch and water-resistant, and the dogs had had very strong, hard feet. John told her these dogs had been capable workers of the small Sutherlandshire cattle.

Mrs Combe considered the kelpies of the Hebrides to be the descendants of a bear-hunting spitz type of Scandinavian origin. (More on spitzes later in the chapter.) She believed (probably alone) that a pure strain of these kelpies existed for centuries in the former Scottish kingdoms of Ross, Cromarty and Sutherlandshire. She says they were commonly coloured black, and black and tan, and occasionally produced chocolate and brown: all colours displayed by Australia's kelpie. Later, when Mrs Combe saw a picture of the legendary kelpie Coil, she thought he looked remarkably like the old original dogs of the Orkney and Shetland Island groups she had never seen, but not like 'John's' Skye 'kelpies'.[4]

There's nothing to see here.

* * *

However, here's an adjunct to the Rutherford mystery out-cross theory: John Rutherford returned to visit his family at Kildonan, Sutherlandshire, in 1865, sixteen years after emigrating. Perhaps by this time Richard Rutherford had sourced the mystery out-cross and introduced it into his collies, and had developed a line of smooth-coated collie dogs his brother had previously requested. It is a very plausible theory, considering that John Rutherford returned to Australia around the same time as Clyde and Lassie – smooth-coated, prick-eared Rutherfords, the first of their kind – reached Australia.

Why would Scotland's premier working-dog breeder go out of his way to develop a smooth-coated dog for cold Scottish conditions? It is quite possible that Richard Rutherford developed his smooth-coated collies specifically for his brother John, and an informal cartel of Scottish-Australian graziers.

The theory continues to grow legs considering the close connections among a small group of Scottish-Australian graziers such as George Robertson, the Tullys, and William Allan of the Elliot and Allan partnership, who were all known to the Rutherfords, and thus gained access to their dogs. The exception to that rule was Jack Gleeson, who came by Robertson's pup (Gleeson's Kelpie) through the luck of the Irish.

It is believed that George Robertson imported the unnamed sire and dam of Gleeson's Kelpie in 1867. But did he?

George Robertson was a meticulous man. Champion Australian shearer and documentary maker Bill Robertson (no relation) found that out for himself when researching *Origins of the Australian Kelpie*.[5] Bill was graciously given full access to George Robertson's records. He writes that George Robertson's record-keeping back to 1854 was impeccable, but significantly Bill Robertson noted that 'among many other purchases Geordie Robertson made over the years, both large and small, there were no dogs listed'.[6] What, no dogs?

Bill Robertson's investigations at Warrock Station were conducted in an attempt to prove his belief that dingo was used to produce the litter that contained Gleeson's Kelpie. That infusion of dingo, as we now know, certainly did not happen. (More on that later.)

But Bill's excellent research into George Robertson's archives may substantiate *another* theory.

Perhaps George Robertson didn't *import* Rutherford dogs. Perhaps he received them directly from John Rutherford of Yarrawonga Station, whom he knew. As there are no records of any dog transactions it's highly likely the dogs were a gift.

Once the smooth-coated Rutherford collies became established in Australia there would have been little call for more expensive importations. Consequently, the Rutherford smooth-coated breeding program in Sutherlandshire probably ceased, having served its purpose. It appears highly likely that only a few litters of the extraordinary short-coated collies were produced. These smooth-coated Rutherford collies are certainly not to be found in Scotland, or anywhere else, today.

* * *

Bert Howard has been looking at yet another but obscure theory that shepherd dogs from North Africa made their way to Britain after the Napoleonic Wars and contributed to the make-up of the Rutherford dogs.

Not likely, mate. Research has found only one possible candidate still existing: the Bedouin shepherd dog. It is a medium-sized dog of distinct pariah ancestry, somewhat similar to the Canaan dog of Israel. It stands about 20 inches (50.8 centimetres) at the withers (shoulders), is smooth-coated and can be virtually any colour, though blue merle is apparently the most prized colour. Desmond Morris also mentions the Bedouin shepherd dog in *Dogs: The Ultimate Dictionary of Over 1,000 Dog Breeds.*[7]

Internet searches reveal only a few sketchy references to it and one grainy, inconclusive photo of a blue merle dog with white points (extremities). One reference claims that the

Bedouin shepherd dog is not fully domesticated, and is extremely stubborn and aggressive, nearly impossible to train and difficult to handle.[8] If that is so, it can't be much of a working dog.

It is highly unlikely that this dog or any similar North African type contributed to the Rutherford-strain collies. The North African theory has nothing to recommend it. That region, arid and largely barren, has never had a reputation for developing stock-working dogs of any note.

* * *

The kelpie's build and durability have even led to suspicions that its short-coated progenitor was to be found in Australia all along.

There is a broad rule of thumb regarding domestic dog appearance and obedience. The more basic a dog is in type – erect ears, long tail, rectangular build – the more basic it is likely to be. Those basic or primitive types are invariably highly intelligent and bold in nature. They are also inclined to be proficient problem-solvers.

The spitz types are all known to be like that. There is no more versatile type in dogdom. Huskies, samoyeds, malamutes and eskimo dogs are draught animals. Karelian bear dogs, Norwegian elkhounds and Japanese akita inus are hunters. The lapphunds are herders, and the Finnish and Swedish spitzes are small-game retrievers and gundogs. The petite German and Japanese spitzes and the tiny Pomeranian are feisty companions. Irrespective of breed, the spitzes have retained certain wolf-like traits. They are usually suspicious of strangers, courageous and independent.

The dingo, while a species rather than a breed of dog, is also a perfect example of this rule.

There have always been comparisons between the kelpie and the dingo, and it is not surprising. The Australian environment forged the dingo, and helped shape the kelpie. The kelpie is a modified but advanced spitz type, and the dingo's forebears were the progenitors of the spitz. Both are rectangular, smooth-coated, prick-eared dogs, built as any sturdy bush marathoner should be: lithe, athletic, muscular. It is the primitive build essential for a dog to survive the trials of the Australian bush. And that is the very point that evades the dingo conspiracy theorists.

There is nothing superfluous about the kelpie. He is basic and spare: the perfect working dog. The dingo has remained unchanged for thousands of years, while the kelpie has taken a long, circuitous route from the other side of the globe to arrive at his type.

It is natural that the kelpie and the dingo share physical similarities. That basic, prototypical build dominates wild environments the world over. It's proof that the founders of the breed did their job. The harsh challenges of the Australian bush profoundly influenced the kelpie's early development. The tough little kelpie emerged from that baptism of fire a century ago, to prove that it is the working sheepdog that succeeds where all the others have failed.

There is a persistent theory that George Robertson, the breeder of Gleeson's Kelpie, had a collie and dingo breeding program in place at Warrock Station, in the same manner as Thomas Hall used the dingo to develop his heelers. Gleeson's Kelpie is said to be the outstanding progeny of that program.

There's not a scrap of evidence that George Robertson even knew much about working dogs, if anything. He was a very proper sort of man. The dingo was the sheepman's despised enemy, declared vermin, and it was illegal to keep one. A dingo breeding program is something he would never have contemplated, even if he knew how to manage one. He certainly left no kennel or breeding diaries or notes.

We know nothing of Gleeson's Kelpie's littermates, and that in itself is significant. They would have more or less possessed kelpie potential, but dog-wise, George Robertson's only contribution to the kelpie story is the dog he lost, not the ones he kept. We know nothing of them.

So, what of Gleeson's Kelpie herself? There are two descriptions of her. A highly respected magistrate, grazier and stockman, Robert MacPherson, who knew Jack Gleeson well, described her thus: 'Kelpie was an ordinary looking well-bred collie of fair size, long haired with long ears, which went up at work. She was brownish-black on back and sides, with some white on breast, and a little about the face and under her jaw, and had tan legs.'[9]

Historian Bert Howard provides what must be considered to be the most reliable description of Gleeson's Kelpie, though it is somewhat inexact. John (Jack) King was involved in the King family's kelpie breeding program on Woolongough Station. Shortly before Jack's death, at a time when the breed was beset by controversy about the origins of Gleeson's Kelpie, he wrote and signed a statutory declaration that stated:

> The kelpie, or old kelpie, known by the name of Gleeson's
> Kelpie, was a black and tan, long-haired, lop-eared,
> medium-sized bitch which had a red tinge in her coat
> when the sun shone on her. When she worked her ears
> went up and down. Her sire and dam were imported from

Scotland by the late George Robertson of Warrock Station, Victoria.[10]

Both descriptions are broadly comparative but self-contradictory in describing Kelpie as having long or lop ears that rose when she was working (aroused). There are four basic ear types: full prick or erect, as seen in almost all the spitz breeds and some of their derivatives; half prick or folded, as seen in the rough collie and many of the terriers; rose ears, as seen in border collies, greyhounds and Staffordshire bull terriers, and full drop as seen in the gun dogs and the scent hounds. Lop or full-drop ears are capable of some expressive movement, but are incapable of rising. It is physically impossible.

There is no question that both descriptions of Gleeson's Kelpie's ears are incorrect. While it is obvious she didn't have full-prick or erect ears, she must have had the mobile, sensitive rose or half-prick type. That is the only non-permanently erect ear type that can be raised to an imperfect full-prick shape when the dog is aroused or excited.

It is possible to produce erect-eared dogs from matings of erect-eared and semi-erect-eared dogs. Therefore, if Gleeson's Kelpie had rose or half-pricked ears it was entirely possible for her to produce erect-eared progeny from erect-eared Rutherford sires.

Much information can be gained from ear shape when divining breed. In attempting to clarify Gleeson's Kelpie's ancestry, her ear type is of crucial importance, if only to clearly indicate what she *wasn't* rather than what she *was*, given the speculation about the early infusion of dingo blood.

Gleeson's Kelpie was *not* part dingo, and here's why. Crossbreeding dingoes with domestic dogs, irrespective of the breed, produces progeny with a broad, dingo-shaped head and erect ears. *Every time.*

The crossing of the dingo with Hall's cur, a half-pricked or rose-eared dog, is a prime example. That mating produced only erect-eared dogs. If Gleeson's Kelpie had a dingo sire as has been claimed she would have had a broad, dingo-type head and general appearance with erect ears.

The widest part of a dingo's anatomy is its head. Dingoes have particularly broad, large skulls. Their teeth are larger than domestic dogs of comparative size, and set more widely apart. Robert MacPherson and Jack King were sheep men and bushmen who would have known an adult dingo when they saw one. They never mentioned that Gleeson's Kelpie was dingo-like. Neither did any other sheep man who saw her or her progeny.

Mr Phil Mylecharane, a highly respected agri-businessman, recorded his personal insights into the backgrounds of Gilbert Elliot and William Allan's Rutherford collies Brutus and Jenny. Mr Mylecharane was a most credible witness, though it appears he may have been somewhat prone to jumping to colour-inspired conclusions:

> In 1870 I went out to Mr Allen, of Geraldra Station,
> to buy flock rams for Goldsb[o]rough Mort and
> Company. When I got there Mr Allen told me the rams
> were out in the paddock, but he would soon get them in
> for me. So saying he opened the yard gate, whistled up
> two smooth, prick-eared black and tan dogs, a male and a
> female, and sent them out into the paddock. In a very
> short time they were back with the rams, and put them
> into the yard.

I never saw dogs work sheep like these two did, and noticing that the bitch had pups, made up my mind that I must have one. So after I bought the rams (I took a big lot of them) I asked Mr Allen where he got those dogs. His answer was that they had just imported them from Scotland from a wonderful working strain there; the dog's name was Brutus and the slut Jenny. [Before the word 'bitch' came into vogue, 'slut' was the common term for a female dog, particularly in the bush.]

I asked for a pup. He told me there was only one left, and he thought I wouldn't like its colour. We went around to see the pups, and he pointed mine out, a little red-coloured one, exactly like a dingo; the rest were black and tan. I thought that it was a dingo. Mr Allen assured me that this was impossible, as the pups were sired on board [during the journey to Australia], and every care taken. He advised me to take the pup, and he would write home to the breeders and see about it.

I took his advice and the pup. The latter turned out a splendid worker. After having him for two years he was stolen from me down at Gooloogong, near Forbes. The next time I saw Mr Allen he told me that the breeders of Brutus and Jenny had written back to say that in nearly every litter they got a similar pup to mine and that they were great workers.[11]

Gilbert Elliot and William Allan had purchased Geraldra from John Kennedy. It is highly likely that it was Kennedy who told them about the Rutherford dogs, since Kennedy had sold another property, Yarrawonga, to John Rutherford.

Mr Mylecharane certainly confirmed that Elliot and Allan's Jenny and Brutus were recently imported, prick-eared, black and tan dogs. Journalist Robert Kaleski (more on him soon) told Tony Parsons Mr Mylecharane had informed him that Brutus and Jenny were Rutherford dogs,[12] though Mylecharane does not state that in the above account.

There are certainly similarities between the kelpie breed and the dingo. But the latter has a distinctive, unmistakeable appearance. Side by side, a smooth-coated pre-kelpie working collie and a dingo of the same colour could not possibly be confused. Cross that pair and stand the progeny next to a similarly coloured dingo and it would be extremely difficult to pick the hybrid. As we'll soon see, a recent Australian study has confirmed that dingo crosses, even specimens several generations old with atypical dingo coat colouring, still look like dingoes. It would have been glaringly obvious if the early kelpie-type dogs were part-dingo, because they would have looked just like dingoes, and they would not have been very reliable workers.

It took Thomas Hall ten years of selective breeding and back-crossing to the cur to achieve a heeler that was not predominantly dingo and still had the working ability he needed. Dingo blood 'worked' in the Hall's heeler because a tough, unsophisticated dog was needed for controlling wild, dangerous cattle. But the qualities required of the heeler were a far cry from the finesse and restraint needed for shepherding lambs.

Thomas Hall was a man with a strong need to produce a serviceable cattle dog. There were no suitable alternatives to be found in New South Wales at the time, so he turned to the dingo – necessity being the mother of all invention. The Hall's

heeler was a well-considered invention, and an enormous leap of faith, and fortunately Thomas Hall landed on his feet.

But the situation thirty years later in southern New South Wales sheep country was entirely different. There was no desperate need to create a super sheepdog. The working collies in Australia in the mid-nineteenth century were certainly doing the job. That isn't to say that good dog men weren't trying to improve them. But no sheep man with his head screwed on would have considered using the dingo, the detested enemy of the wool industry, to *create* a sheepdog.

A 2016 study by a team of Australian scientists conclusively kills off the claim that a dingo sired Gleeson's Kelpie.[13] That study found that the dingo skull shape remains unchanged by crossbreeding with domestic dogs:

> 'We know that cross breeding has an effect on the dingo gene pool but what we didn't know until now is whether cross breeding changes the dingo skull,' said study lead author Dr William Parr, Postdoctoral Research Fellow at UNSW [University of New South Wales] Medicine's Surgical and Orthopaedic Research Laboratory.
> 'This study has shown us that the dingo skull shape, which in part determines feeding ability, is more dominant than dog skull shapes,' Dr Parr said.[14]

The research team used medical CT (computed tomography) scanners to make three-dimensional models of the skulls of dingoes, domestic dogs and hybrids. They found that hybrid skulls were indistinguishable from those of the dingo, either with the naked eye or statistically. The researchers think that the dominance of the dingo skull shape is most likely due to the fact that recessive traits are fixed in dogs. They believe selective breeding of domestic dogs – often targeting recessive characteristics – has narrowed their gene pools:

> 'What we found was a strong convergence on the dingo-type shape,' [co-author Dr Michael Letnic] said.
>
> 'There was a strong tendency for animals to look like dingoes regardless of whether they were a pure dingo or had been cross-bred …
>
> 'Those characteristics becoming evident in domestic dogs – things like floppy ears or particular shapes of dogs or hair colours – often is a result of people targeting for specific mutations that are often deleterious – they're not things that are advantageous in the wild,' Michael explained.
>
> 'Dingoes have been living in the wild for thousands of years, and they've had a very strong selection for the characteristics that are required for animals to survive in the wild,' he said. 'Those wild-type characteristic[s] also appear to be genetically dominant.'[15]

* * *

So the question of whether Gleeson's Kelpie or any of the other early kelpie progenitors have dingo blood can now be laid to rest – though that is not to say that there is no dingo blood in some kelpie lines. There is, because there were deliberate infusions of dingo blood into some kelpie strains during the mid-twentieth century, though there cannot be much remaining in the twenty-first century.

It is probably reasonable to assume that dingo-meddling with kelpies was occurring from the early days of the breed.

Perhaps there were unrecorded kelpie-dingo litters produced that may have started the dingo origin theories that journalist Robert Kaleski would later claim were current. We will never know, but some twentieth-century breeders, prominent men, thought there was some merit in introducing dingo blood.

DNA testing has confirmed what some kelpie men have already admitted: there is dingo blood present in some kelpies today. The testing was facilitated by Bill Robertson, the author of *Origins of the Australian Kelpie*, though it was a sampling of only seven dogs and certainly not representative of the breed as a whole. The genetic testing identified mainland and Fraser Island dingoes as being present in the tested samples. The introduction of Fraser Island dingo blood probably occurred well after the kelpie emerged as a distinct breed and made their way onto Queensland sheep stations in the early 1900s. There has never been wool growing on Fraser Island, though there were plenty of sheep on the mainland just across the Great Sandy Strait to the west. The Fraser Island dingo infusion could have occurred 100 years ago or five years ago. We don't even know if it was an intentional mating.

Accidental matings between kelpies and dingoes could always occur in sheep-raising districts where dingoes were present. The infusion of dingo doesn't appear to have been of any appreciable benefit to the kelpie. So much back-crossing to the kelpie was necessary before the dogs showed any reasonable gathering aptitude that it made the effort hardly worthwhile. The kelpie is such an adaptable, superior breed that it never needed the dingo to make a hardy Australian sheepdog of it. The dingo meddling occurred at a time when pure kelpies were consistently outperforming any other working sheepdog in tough conditions. The introduction of dingo blood has never made a better working dog of the kelpie. Just another case of trying to fix a working breed that clearly wasn't broken.

In a letter dated 19 July 1948, Frank Scanlon, a great kelpie man from Quirindi in northern New South Wales, told Tony Parsons that he had no doubt the kelpie originated in Scotland and that the then recent infusions of dingo had done no good for the breed:

> I have gone as far as crossing a dingo bitch (pure dingo) and a particularly good kelpie and the dog I kept from this cross was the most useless, brainless brute one would ever call a dog. I passed him on to Goodfellow (he at that time was bitten with that dingo stuff). The kelpie got his working ability from Scotland and not from any dingo mong. Also, all colours including yellow (dingo) colour, have come from Scotland.[16]

Frank Scanlon obviously wasn't one to be fooled by colour similarities. A border collie man, Dr R B Kelley, writing in *Country Life* on 16 June 1950, stated that he knew of several infusions of dingo with the kelpie, and that the progeny were nowhere near as useful as pure kelpies until they had diluted the dingo blood to 1/16 or even 1/32.

'On the other hand,' Scanlon wrote, 'Jack Goodfellow, who worked his two excellent dogs, Don, and Nap, so successfully a few years ago in Sydney and elsewhere, believed a dash of dingo blood was advantageous.'

So Jack Goodfellow, a great kelpie man in his own right, had faith in the dingo out-cross, but a dog with just a dash of dingo in it.

He told Tony Parsons that some dogs he bred from the kelpie and dingo, even the first cross, were quite good dogs for driving mobs and very hardy, with good feet. The trouble was, for a few generations there were a lot of duds in each litter. 'You may get two that will work and make champions and four that will never look at sheep, out of a litter of six. Strangely enough, they are not sheep killers. You will get an odd one that will bite as a pup, but they are very easy to train not to.'[17]

Jack Goodfellow was obviously keen on the dingo cross, but his assertion that two-thirds of a crossbred litter were useless as workers is hardly a glowing endorsement for a type he thought was superior to the pure kelpie.

Goodfellow was not only a dingo x kelpie adherent, but he was also a Scottish working collie denier. He wrote to Tony Parsons: 'Now a word re the joke about the Scots breeding kelpies … or the breed originating from Scotland. If you have studied dogs at all, you can go over all the breeds of dogs in Scotland and there was never a combination of dogs in the land that could breed a working kelpie … or any other land for that matter where there was no Australian dingo.'[18]

That Jack Goodfellow was a better kelpie man than he was a kelpie historian goes without saying. He could tell us what he thought the kelpie wasn't, but he was unable to enlighten us on what he thought the kelpie was. Besides part dingo.

* * *

In one respect Jack Goodfellow was partially and accidentally right: the smooth-coated Rutherford collie of Scotland was not to be found in Scotland. It does have an existing close relation, but there is no point looking in Scotland for it. It is the kelpie. However, from the kelpie's primitive conformation it is reasonable to conclude that the Rutherford mystery out-cross was a spitz-derived working collie type. The smoking gun is most likely held by one of the early German sheepdogs, as originally suggested by Tony Parsons, or a dog very much like it.

Due to the passage of time and the loss of so many breeds and types, and considering the dog's breakfast that was the mainstream dissimulation of our working dog origins, that conclusion, an educated guess, is the best we can manage at this stage.

CHAPTER 11

Cattle Dog and Kelpie 'Myth-information'

Some breeds are so ancient it is impossible to figure out how they developed. As we have seen, even the origins of more recent breeds can be difficult to establish. Certain aspects of the kelpie's history are likely to remain shrouded in mystery. Yet in the early 1900s determining that Australia's heelers were derived from an English cattle-droving dog should have been reasonably straightforward.

As the nineteenth century approached the twentieth, wool was booming, and Australians had larger disposable incomes and were enjoying a high quality of life. Old breeds like the kangaroo dog were declining in popularity, and the pedigreed companion dog of almost any new available breed was fast becoming the next big thing. In response to Australia's increasing interest, it was inevitable that commentators would emerge to service the growing demand for information on dog breeds and their origins.

The first was Walter Beilby, a prominent Melbourne-based dog judge, and the father of dog showing in Australia. Beilby was an Englishman and he embodied the very image of imperial dog-show chauvinism. He was a bigoted, pedigree dog snob, but there is no question he knew his dogs, having been heavily involved in the hobby in England before emigrating.

To meet the demand for pedigree-dog-related information, Beilby published *The Dog in Australasia* in 1897. It was a comprehensive overview of the most popular dogs in Australia at the time, featuring around sixty breeds, including standards and detailed pedigrees for some of the better specimens of most breeds.

Beilby was the authoritative voice of the pedigree-dog world in Australia, but he had no interest in the Australian native working breeds, whose stronghold was New South Wales. *The Dog in Australasia* was acerbic in its treatment of them. Beilby lambasted the little Australian terrier and ignored the 'cattle dogs', as the Hall's heelers had become known. He barely mentioned the kelpie, gave the kangaroo dog short shrift, and wished rapid extinction upon the dingo. It left the native working breeds without a voice and champion. But not for long.

* * *

Robert Lucian Kaleski was born in Sydney in 1877 – the same year as Tasmania's ultra-punishment prison camp at Port Arthur closed, and just seven years after Thomas Hall's passing. He grew up at a time when Australia's great retrospective love affair with colonial times and the bush had begun.

No country in the British Empire, let alone the new world, developed such a dependence on stock-working dogs as Australia. Banjo Paterson and Henry Lawson were the great literary entertainers of the day. The adventures of Lawson's 'The Cattle Dog's Death' (1893) and 'The Loaded Dog' (1901)

brought the bush dogs into the lives of urban Australians and fired the public's imagination.

As a teenager in Sydney, Kaleski discovered Hall's heelers. These hard-as-nails cattle-wranglers had a boisterous, enigmatic air about them: half dog, half dingo, coloured like the iconic red and blue kangaroos and the red earth and blue sky of the outback. No one in Sydney knew much about them, though they were certainly familiar through the Halls' cattle business.

Robert Kaleski was astute enough to recognise an opportunity when he saw one. Embracing a 'wild colonial' writing style, he began submitting articles about working cattle dogs to Sydney newspapers. He was certainly a capable and entertaining writer. He needed to be, because he had nothing to recommend him as an expert on Australian working dogs. He was an urban enthusiast and dog exhibitor who had had nothing to do with the development or maintenance of working strains of heelers or kelpies. But he had ability, ambition, the good fortune of being in the right place at the right time, and the greatest of all good fortunes: a clueless audience.

With more front than Gowings, and no definitive breed histories to moderate his bizarre biological and historical notions, Kaleski created a façade of credibility through a contrived image of the dog-wise, laconic bushman. Editors were desperate to satisfy the public's curiosity and granted him plenty of copy space. And the readership lapped it up. They loved Kaleski's rip-roaring yarns, and Kaleski sold papers. He was to become one of the most prolific and popular columnists in Australia, and for fifty years he contributed dog-related articles to *The Bookfellow*, the *Sydney Mail*, the *Sydney Morning*

Herald, *The Bulletin*, *The Land*, *Walkabout*, *The Worker* and the *American Kennel Gazette*.

When it came to Australian dogs, Kaleski was your man. He was passionately devoted to the kelpie and the tailed Hall's heeler. He almost singlehandedly brought the working breeds to national prominence, and naturally, just about everyone took him seriously, but virtually everything he wrote about the origins of the Australian working dog was the product of his imagination. And no one, other than a couple of exasperated scientists, and his show-ring enemies (there were a few), ever took him to task. Always half cocked, prickly and emphatic, but never wrong, he didn't scruple to publicly berate anyone – including the scientists – who questioned his bizarre claims.

Kaleski was very well connected. The poet Dame Mary Gilmore, a fellow *Bulletin* contributor, was a fan, and they were big mates. In her 1922 book of prose *Hound of the Road* she wrote: 'The idyll of the hawk is written; the song of the nightingale rises from the printed page, the Swiss loved one dog and set him up in stone. But who has written our dog? Kaleski? Kaleski wrote dogs, not the dog.'[1] Whatever that all means.

In 1914, Kaleski published *Australian Barkers and Biters*,[2] unquestionably the most influential book ever written on Australian dogs. It mostly comprised articles previously published in his popular articles in the *Sydney Mail*. Hailed as the definitive treatise on Australia's dogs, *Australian Barkers and Biters* was enthusiastically received over several editions.

Read today, *Australian Barkers and Biters* is cringeworthy. Kaleski's breed knowledge was extremely limited and often incorrect, and his technical knowledge abysmal. He considered himself a serious amateur scientist and historian, but he had no head whatever for conventional science or history, being too given to wild, unsupportable theories.

However, *Australian Barkers and Biters* was spectacularly successful in propagating Kaleski's bizarre theories and misinformation about Australia's working dogs, imported dogs, and even the dingo.

* * *

When the kelpies hit town in 1896, Robert Kaleski wasted no time in jumping aboard the bandwagon. The pioneer kelpie men of the bush were all Kaleski contemporaries, and a lot of them came to Sydney every year for the sheepdog trials and the Sydney Royal Show. As a working journalist, Kaleski would have known and spoken to all the major working-kelpie players. He was perfectly placed to glean the facts of the kelpie's origins, to the extent to which they were known. Yet Kaleski's stubborn nature and unfathomable biological notions made a joke of the kelpie's origins.

He said most people thought the kelpie was a cross between the smooth collie and the dingo – the same breeding he had claimed created the Hall's heeler. However, he surprisingly rejected that incorrect theory and maintained that men from the Scottish–English border country had told him that the kelpie originated from a cross between a fox and a black smooth-coated collie created by a gypsy – for poaching, of all things – 100 years before it appeared in Australia.

Kaleski was talking through his hat, though as we've already seen, the mythical dingo out-cross theory has always hovered above stories of the kelpie's origins. Sometimes, Kaleski wrote, you have a red pup, just like a fox in a litter, and other times a blue, which he claimed was a relic of the Highland collies. How blue Highland collies came to be involved he did not state, but he gives his singular opinion of their worth based on their colour: 'The pups are picked in the same way as Cattle-pups for markings, color and hardiness … For color, black and tan is the best, then the blues and lastly the red or sandy.'[3]

Any sheep man who selected a working pup based on markings and colour would have been a fool, and hardiness cannot be determined in infant puppies. Credibility with salt-of-the-earth types is only ever earned. Kaleski had none with the genuine kelpie men of the time. It is no surprise they were hugely bemused by his inexplicable, senseless ramblings.

Kaleski insisted all the kelpies threw to the fox, except the all-black ones, which threw to the collie, and were therefore a separate breed, which he called the Barb. The original Barb was just a black-coloured kelpie, named after the two-time Sydney Cup and 1866 Melbourne Cup winner, The Barb, a black stallion with a nasty reputation. Other than its colour, there was nothing different about the black kelpie.

In *Australian Barkers and Biters*, he informed the reader that the kelpie type was a small, smooth, prick-eared collie, with a dash of fox in the black and tan and the blue, and a very strong dash of fox in the red. The Barb type was a small, smooth black collie, but with prick ears. A prick-eared dog was, he said, useless in Scotland because the snow got into its ears and sent it deaf.[4] He continued:

> Many people in this country do not know the difference
> between a Barb and a Kelpie; many bushmen, even, think
> they are the same, but anyone working them soon finds out

the difference. The Kelpie is shy and silent and fond of working his sheep very wide; the Barb, on the other hand, likes to get close to his work, bark when it suits him and nip the sheep a little if he thinks they want it. The reason for the difference is very simple: the Kelpie throws back to the Fox, and the Barb to the old Border Smooth Collie …

So the Barb's origin (throwing back to his Collie grandparents) explains why he can handle the English sheep, also why he is not sensitive, like the Kelpie. He has inherited the instinct of the Collie, which has been so long with the shepherd that it has learned his ways and to a certain extent has developed itself up to his brains. He knows just about what a man is and bears himself accordingly. When work needs doing, he does it; if there is none to do, he is quite satisfied to eat and sleep till there is. Therefore, unlike the Kelpie, with whom work is a passion, the Barb never exerts himself more than is needful. Similarly, if he finds he has a bad master it is not long before he finds a better one. He reasons, as compared with the ordinary dog's instinct.[5]

If that wasn't enough to convince the working-kelpie men and women that Kaleski was completely deranged, or publicly practising upon them and their dogs, then nothing would. Tony Parsons cites instances where kelpie men familiar with the Kings and the great kelpie man Jack Quinn indicated to him that they were unwilling to cooperate with Kaleski and had little time for him.[6] As far as the working kelpie went, Robert Kaleski found himself an outsider looking in. His insistence that the kelpie and Barb were two separate breeds and his bizarre fox-progenitor claims marked the point when his unchallenged opinions and reputation began to be called into question.

* * *

So embedded are Kaleski's myths that despite the best efforts of credible twentieth-century cattle-dog researchers and historians such as Bert Howard and Noreen Clark, authors, breeders, breed clubs, breed and pet advisory websites and even some trusting scientists continue to perpetuate his origin myths concerning the kelpie, and particularly the cattle dog.

Kaleski lived just 150 miles from Dartbrook in the Hunter Valley, less than 50 miles from the Halls of the Hawkesbury, and less than 15 miles from their Liberty Plains (present-day Auburn) holding paddocks. Even though Hall's heelers emerged just thirty-seven years before his birth and worked around Sydney and its districts, Kaleski was woefully ignorant of their origins, and he never got it right. He obviously felt no need to research. But all he needed to do was ask the right people.

The Hall beef cattle empire had wound up by the time Kaleski began writing about the heelers, yet there were still plenty of accessible Hall contemporaries familiar with the development of the blue and red dogs. Admittedly, Thomas Hall's breeding program was a private affair, but New South Wales was a small place in the mid-nineteenth century and was little larger by the turn of the twentieth.

In 1897, Kaleski first claimed that the development of the Hall's heeler occurred around 1860, when: 'A Mr Hall or Wall of Muswellbrook imported the blue-grey Welsh merle for working cattle, but finding they were unsuitable on account of barking too much, crossed them with the dingo and founded the present variety, which, by selection and careful breeding, became a distinct breed and throws true to type [produce offspring that all look like the same breed].'[7]

He later changed his story for *Australian Barkers and Biters*:

So Mr Hall, of Muswellbrook, imported some Blue
Smooth Highland Collies or Merles, called by ignorant
people Welsh Heelers. These were a lot better than the
common Collie, but still had some of the barking-at-the-
head business in them. So they crossed on the Dingo.
(The early settlers had a great respect for that animal.)
This turned out all right: the pups came blue-speckled or
red-speckled as the Merle or Dingo was stronger in them.
Instead of yapping like the Collie, they had the old
Dingo style of creeping up behind and biting. It is the
dingo which makes the Blue and Red Speckles the best
Cattle-dogs in Australia to-day. These Speckled Heelers
are like a small, thick-set Dingo to look at, except in
color; if you met one in the bush you would shoot it for
its scalp.[8]

As we have seen, little of that is true. Kaleski's claim that a gathering-type sheepdog was crossed with the dingo to create a dog that could stand up to and out-bully horned wild cattle is just absurd. But no one questioned it for decades.

It was Kaleski who started the furphy that the Hall's heeler was developed in Muswellbrook. Muswellbrook has no claim as the birthplace of Hall's heeler. Hall's property Dartbrook was adjacent to the current town of Aberdeen, 20 kilometres north of Muswellbrook. But, exploiting the

Above: Neville Butler with two Timmins biters, circa 1900, showing some variation in colouring and markings. Right: A Hall's heeler, circa 1890. Note how well the build is suited to a dog required to cover many kilometres of hard running every day. (Both photographs courtesy of Bert Howard)

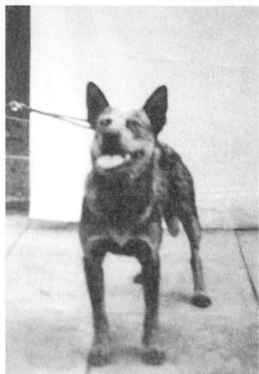

popular Kaleski myth, Muswellbrook claimed the honour and commissioned a grotesque statue of a 'blue heeler' that was as bereft of authenticity as Muswellbrook's claim to the Hall's heeler's development. In 2015 the old statue was mercifully put out of its misery and a new one installed. It's lightyears better than the old one, but it is built as solidly as a mastiff and does not replicate the dog that was developed at Dartbrook. Just sayin'.

Kaleski later *again* reinvented his invented heeler history when he made the bizarre claim that breeders had decided to put Dalmatian blood into the Hall's heeler to give it 'a love of horses and guarding instinct'. He also claimed that kelpie blood had been introduced to the early Hall's heeler because it was 'wanting in working ability'.[9]

The legacy of these and many of Kaleski's other claims is that for over a century, people around the world believed that these alleged crossings influenced the breed as a whole – that all cattle dogs have been bred from the smooth collie, the dingo, the Dalmatian and the kelpie.

Robert Kaleski messed up his heeler origins because he was unaware of Hall's cur and its vital contribution to the breed. But to be fair, it wasn't just Robert Kaleski who went into print to butcher the cattle dog's origins. There were some very silly versions getting around, but they didn't gain the broad acceptance that Kaleski's popular theories did. Kaleski's erroneous version of the origins of the Hall's heeler was the official cattle-dog entry in the *Australian Encyclopaedia*,[10] and his myths and misinformation went mainstream into libraries, homes and schools until the last edition in 1996. Dog cred, it appears, was all in the telling, and Kaleski was a master of the tall story.

* * *

Despite its isolated development and a century of populist misinformation, as we saw in Chapter 8, the origins of the Hall's heeler and the subsequent divergence of the Timmins biter are now clear. While the identity of the British dog that contributed to the Hall's heeler was open to distortion, no one has ever doubted that the native progenitor of those mighty breeds is the dingo. In fact, for cattle-dog enthusiasts and other Australians in the know, the dingo's contribution has always been something to be proud of. It is the world's unique wild-dog redemption story.

It's the greatest pity that the story wasn't told until the last years of the twentieth century. And it wouldn't be a dog man, but one of Australia's great dog women who smelled a rat and finally created the impetus for the real story to be told.

Bernie and Berenice Walters of the famed Wooleston Kennels and the Australian Native Dog Training Society of Bargo, New South Wales, were very influential cattle-dog breeders for several decades. They originally came from Moree and were familiar with the working Hall's heeler types, both tailed and stumpy; as mentioned, Moree was formerly part of Weebollabolla Station, one of the Hall family properties. Bernie and Berenice knew their cattle dogs. And later, their dingoes.

Their great champion, Wooleston Blue Jack, born in 1954, was the most influential show-strain cattle dog of the post-war era. Wooleston Blue Jack provided the means for Bernie and Berenice to produce dogs capable of both winning in the show ring and working cattle. If any show strain came close to approximating the original Hall's heeler,

it was the highly regarded Wooleston dogs. For decades, Bernie and Berenice were able to produce workable cattle dogs of good type.[11]

Berenice Walters had long chafed at many of Kaleski's claims and theories. She knew enough about cattle dogs to know a tall story when she heard one, and she was frustrated that the true origins of the Hall's heelers were unknown to even the most dedicated owners, exhibitors, stockmen and breeders.

However, even she didn't know the full story. It was a meeting with Bert Howard, that proved to be the turning point in uncovering the truth about the origin of the Hall's heeler, and eventually, the kelpie.

Bert Howard has a close connection with the Hall family: his late wife, Beryl, was a Hall descendant. In July 1975, he paid a visit to Berenice Walters at Wooleston Kennels to buy a puppy. At that time, by sheer coincidence, Bert was researching the Hall family history and George Hall's journey to New South Wales in the *Coromandel*, as part of his contribution to a book, *Over-Halling the Colony*, being produced by Hall family descendants.[12]

On learning about Bert's historical research into the family that produced the heeler, Mrs Walters raised her concerns with him regarding Kaleski's Hall's heeler history. Bert's Hall family research subsequently expanded to include the origins of the heeler.

It took three decades of scouring British, colonial and Hall family records, and collaborating with historians and family members in Australia and Britain, to uncover the real story of the Hall's heeler. The research covered the Halls' original properties in the Hawkesbury region, their expansion into the Hunter Valley and north into Queensland, the bobtail, the red bobtail, the Timmins biter and the Timmins family.

Having established the true account of the development of the Hall's heeler by the mid-1990s, Bert decided that the origins of the kelpie were also worthy of further examination. Again, his research led him to prominent and reliable working-kelpie men such as the late Mike Donelon, the late Jack Goodfellow, the late Jack Body and Tony Parsons. Ably assisted by researchers in Australia and Great Britain, Bert Howard has revealed the kelpie's true origins – though some details, such as the kelpie's mystery progenitor, have eluded the most thorough and persistent research.

Bert Howard is a modest man who has sought no reward or recognition for his extensive labours. He has gone about his research with only one aim: to reveal the truth about the origins of Australia's two prominent home-grown working breeds. By his own admission he is no dog man, but his dispassionate, thorough, even-handed research has quietly dissipated over a century's misinformation. And in doing that, he has done Australian colonial and working-dog history an enormous service.

* * *

Robert Kaleski revelled in his high profile. It is his name that most people connect with the Australian cattle dogs, though in truth he was little more than a very effective publicist who did little to acknowledge the true origins of the breed he so loved.

Yet despite his shortcomings, he did much good. In his own odd way, he was a great champion of the working breeds

he loved. He did more than anyone to raise general awareness of the cattle dogs and the kelpie.

In her excellent book *A Dog Called Blue*, Australian cattle-dog historian Noreen Clark says that 'his services to Australian working dogs, as a publicist, were monumental'.[13]

Tony Parsons wrote that for more than fifty years there was hardly a major publication that did not carry stories by Kaleski about the kelpie and the cattle dog:

> He was an institution, and although scientists disagreed with many of his theories, he occupied an almost unchallenged position as an authority on Australia's working dogs. Much of what Kaleski wrote (and did) was of great value; the difficulty lies in deciding where he was right and where he was wrong. I knew Kaleski personally and I appreciated then, as I do now, how much he loved our working dogs; without him, the kelpie and the cattle dog would have lacked an effective voice.[14]

As time passed, Robert Kaleski lost favour with a great many dog fanciers as his dated and essentially incorrect information became exposed to the trials of the passage of time. For a time during his dotage, Kaleski took Tony Parsons under his wing and considered him to be something of a protégé, but the irreconcilable differences in their opinions saw the gulf widen between them. When Parsons, a genuine dog man and a rural journalist, publicly challenged some of Kaleski's more bizarre claims, the relationship ended.

Irrespective of the widespread acceptance of Robert Kaleski's 'myth-information', within the Australian and international dog world, Bert Howard has proven beyond doubt that it is Thomas Hall alone who deserves recognition as the founding father of Australia's remarkable cattle dogs.

The last word on that score belongs to the late Mrs Berenice Walters: 'It's a national disgrace for Mr Thomas Hall to be almost unknown while a lot of benchies [dog-show exhibitors] round Sydney claim the glory of founding the breed.'[15]

Amen.

CHAPTER 12

Cattle Dogs and Kelpies in the Show Ring and Suburbs

While interest in dog breeds was growing as the twentieth century replaced the nineteenth, dog shows were also becoming enormously popular in the cities. There was burgeoning interest in introducing Australia's native working breeds to the show ring.

The basic principle of dog showing is to judge a dog's conformation – that is, how closely every aspect and feature of the dog conforms to the breed standard, a detailed description of the perfect – and unattainable – specimen. Every show features a range of age and status classes, and dogs compete against each other to be the best in that class. The class winners then compete to win the best of breed, and the best of breed winners compete against other best of breed winners to win best in the relative group – terriers (vermin killers and some fighting dogs), gun dogs (Labradors, pointers, spaniels), hounds (you should know a few of these by now), and working (and these), toys (the noisy little ornaments), non-sporting (an unclassifiable group that includes bulldogs, chow chows, Dalmatians, poodles and shih tzus), and the utility breeds (often human and stock guards like mastiffs, Rottweilers, and maremmas, or draught breeds like huskies and malamutes).

The best in group winners compete to determine the best dog in show.

In theory it all sounds simple, but judging is a very subjective thing, based on the individual judge's interpretation of the standard, and his or her impression of what is being presented for appraisal. Dog shows have always been a very emotional pastime, rife with disappointments, jealousies, and accusations of prejudice and favouritism. But in the late nineteenth century they were enormously popular, and Sydney-based enthusiasts were determined to make pedigree show dogs out of the native working breeds.

Robert Kaleski was certainly the pivotal figure in the early Sydney dog-show scene. He was frequently outraged when his dogs weren't put up (awarded first place) at dog shows. In 1897, there was still no cattle-dog breed standard, so Kaleski did something about it. He wrote one himself. Just like that! His new standard identified only the blue-speckled, tailed types. For reasons best known only to him, he did not include teeth and feet in the standard, properties one would think essential for a cattle-heeling, long-distance droving dog. Nor did he acknowledge the stumpy-tailed dogs.

The truth was, like all the other Sydney cattle-dog exhibitors, Kaleski was vastly inexperienced with the breed. Kaleski admitted he only got into pure-bred blue-speckled dogs in 1893.[1] He was a lad of sixteen then. Within just four years, at the age of twenty, he had not only contrived to become the self-appointed authority on the breed, but he had also written the standard for it, and declared it could not be bettered. Kaleski might have been short on experience, but he was long on confidence and self-belief.

The Cattle and Sheepdog Club of New South Wales accepted Kaleski's standard in 1903. Coincidentally, Kaleski had started the club and was its honorary secretary. The Kennel Club of New South Wales also accepted his standard – for a time.

Unsurprisingly, the Kaleski Standard – as it became known – mirrored his image of the ideal cattle dog. Instead of spelling the end of his problem with dodgy judging decisions, it proved to be the catalyst for a lifetime's dissatisfaction with the direction taken by the show strains that became known as the cattle dog. Kaleski appeared to be the only person who believed his standard could never be improved. But the times they were a'changing, and so were breed standards. Kaleski railed against the changes to his standard and maintained the rage for the rest of his long life.

Kaleski at least tried to maintain his interpretation of the original Hall type. Most of the other show cattle dogs lived in suburban backyards. Kaleski worked his dogs on his small farm at Liverpool, and believed that for a dog in the show ring to be judged as fitting the standard, it must be built to do the job it was created for. Fair enough, too – except that the Hall's heeler's job had gone the same way as the mid-nineteenth century, and moving a handful of quiet cows about on a small holding is hardly wild cattle work.

Kaleski championed the blue-speckled dogs but inexplicably ignored the red speckle, a dog more reminiscent of the dingo than the blue. That prejudice would be adopted by the majority of cattle-dog fanciers and would prove hard to eliminate, particularly in Brisbane, where the show fraternity favoured Sydney-bred dogs. Blue, solid or speckled, has always been the predominant, and consequently most popular colour.

The show ring, no matter what breed enthusiasts claim, has vastly different expectations from those of the working fraternity. The show ring is all about looks. The dog-show scene has always been criticised for producing beauty-pageant contestants, and little else.

Fashion wins ribbons. Trends in judges' preferences and interpretations cause breeders to change the shape of their dogs. Usually the changes are gradual. This phenomenon is called 'type creep'.

It certainly happened to the cattle dog. In many show strains the (tailed) cattle dog became a stockier, more solidly built, lower-to-the-ground dog, straighter in the stifles (the angle of the hind legs), and vastly divergent from the mile-eating marathoners Thomas Hall created. But still a mighty dog in anyone's book – other than Robert Kaleski's.

There is a reasonable suspicion that the more solid, squat show strain was produced by the introduction of the English bull terrier. Alan Forbes was a prominent New South Wales dog judge who wrote *The Interpretation of the Australian Cattle Dog Breed* for the Royal Australian Agricultural Society Kennel Club. He later stated that 'although there is some disagreement as to the actual breeds used, it is thought that the cattle dog developed chiefly from a crossbreeding between a dingo and a Blue Merle collie, with a later injection of Bull Terrier Blood'.[2] Even officials believed Kaleski's blue merle collie furphy. Everyone did. But Forbes was probably right about the injection of bull-terrier blood into the show lines.

A hundred years ago, the bull terrier was not such an exaggerated dog as it is today. It was lighter in build, and its now football-shaped head was not as severely pronounced. It is a single-coated dog and quite straight in the stifles. Straight

stifles are not conducive to a long-distance running breed. Cattle-dog show strains have been plagued by hind legs that are too straight; that would not have come from either the dingo or the cur.

Many early show lines displayed signs that bull terrier blood had been used. Kaleski's own Thornhill Tiger was a particular case in point. He was a single-coated dog with a poorly defined stop (the step from the forehead to the muzzle) and small, pointy, bull-terrier-type ears.

The Hall's heeler, in the Sydney show rings was fast being superseded merely because of preferences in appearance. Or the Sydney cattle-dog exhibitors, the 'benchies', thought they could improve upon Thomas Hall's heeler.

* * *

Type, or build, was only one of the two missing parts of the show-strain working dogs. The other, most important missing part, for a working dog, is the bit that lies between a dog's ears. And once that part goes missing, it is impossible to retrieve it.

The loss of the ability to perform the role for which they were initially bred is an interesting phenomenon in show or pet types. Generally speaking, stock-working dogs appear to be the most affected, though gun dogs can also completely lose the ability to work cooperatively with a hunter. Both stock and gun-dog work are highly sophisticated adaptations of the hunting instinct. Breeds that employ more basic or instinctive drives, or work styles, seem to be less affected by the neutralising role of a companion dog. Terriers retain much of their original vermin-killing drive, while both the scent- and sight-hound breeds (beagles and greyhounds, for example) seem to have mostly retained the instinct to pursue game and neither beagles and greyhounds nor their close relations, can be trusted when not on lead, so strong are their hunting instincts.

The loss of working ability occurs because working instincts or instinctive drives will wither if not constantly used. Use it or lose it. It only takes a couple of generations of breeding working types who do not work before non-working progeny are produced. The progeny may look identical to a working version of the same breed, but the innate working abilities are lost, while quite often the basic primal urge to chase is not. Around stock, a non-working-strain dog is just a nuisance, or worse.

The loss of working ability is simply the result of a dog's adaptation to its environment and living conditions. If it doesn't need to work, it quickly adjusts to being a pet or show dog, or whatever other purpose it is kept for.

A working drive that makes for an impatient, restless, hyperactive dog is not wanted in show exhibits. Irrespective of breed, no matter what breed standards claim. Every show dog must be a patient, compliant animal that can stand statue-still, tolerates handling and examination by judges, puts up with grooming and being very close to other dogs, and obeys all commands given to it. Working traits in show lines are not desirable, so naturally, breeders select more showing-inclined specimens and breed from them.

Dogs have been manipulated into an astounding array of shapes, sizes and colours. The jobs for which they have been created vary from minuscule companions such as chihuahuas and papillons to wolf- and bear-hunters like the Irish wolfhound and the Karelian bear dog. The one thing every breed has in common is no matter what job they perform, they

are all willing to accept human companionship. Most of them never lose the instinct to protect their territory and people. The dog is a truly remarkable creature.

* * *

Unlike the dog-show fraternity, which focused on appearance and conformity, the bush only cared about ability. If it was built for the job and had enough between the ears, it was bred with whatever else could do the job, tail or no tail. Bush pragmatism mated with necessity. The cattle dogs – both varieties – were interbred with each other and crossbred with other breeds to suit the prevailing work and working conditions. Cattle dogs, or heelers as they are commonly called, became varied in type from region to region and breeder to breeder.

They were now a labour-saving aid rather than a wild cattle-controlling necessity. Crossing heelers with kelpies was one compromise for some progressive graziers, not always to improve working ability, but to modify working style to suit quieter cattle. Later in the twentieth century, kelpies and even border collies worked quiet stud and dairy-farm cattle. Beef producers valued the benefits of raising quieter, polled cattle. Many preferred the more mobile, less forceful working dog, and some needed no dogs at all, but the heeler was a dog capable of re-inventing itself.

With cattle heeling redundancy staring them in the face, the Australian heelers rode barking into the early twentieth century on the back of the motor lorry, and found a new calling. Cattle work became primarily vehicle-based for many beef producers and those who still used heelers found their dogs appointed themselves truck guardians when they weren't working stock, many virtually living on the truck trays. No breed has ever developed such an attachment to the motor vehicle.

As the century progressed, working men from all walks of life and all corners of the country had heelers riding shotgun on the backs of their trucks and utes. The red or blue dog on a ute would became another iconic Australian working dog image, though one best viewed from a sensible distance.

Some heelers worked and lived solely as truck dogs, loading and unloading cattle and traveling in cages under the tray or trailer. But it's as ute dogs that many Australians identify heelers nowadays because they have seen them in no other role.

The apples didn't fall very far from the tree. The Australian heelers are the closest we ever got, or ever will get, to making a dog of the dingo, and they are so primal in some of their instinctive behaviours it's surprising they're not inveterate howlers.

Yet the cur blood has made the heeler obedient, willing to please, and devoted to their family, particularly children. Their faithfulness, courage and protectiveness are as legendary as their bite-first-ask-questions-later reputation, and the stories of their faithful heroics are as common as those of their felonious outrages.

Take, for instance, the heeler that allowed the meter-reader into the yard but wouldn't let him out. The heeler that prevented a toddler from walking into a dam. The heeler that tore the tyre off the postie's bike. Or the heeler that woke its family and saved them from a house fire. The truth is, Australia's heeler, tailed or tailless, show-strain, backyard- or bush-bred, is every one of those dogs, and more.

Raised and managed right, the heelers are tremendous dogs, but only for people with plenty of dog time, a willingness to learn, good fences and a lot of dog sense. Australia's pirate-patched, bandit-masked heelers are the true ockers of dogdom – diamonds in the rough, faithful to the end, and as game as Ned Kelly.

* * *

The Great Depression of the 1930s severely limited most social and sporting activities in Australia, dog shows included. Dog showing resumed shortly after the Second World War, and established the Australian cattle dog as one of the most popular working breeds exhibited in Australia.

By this time, the, small dingo-looking working Hall's heeler type in the show ring was no more and the new show type, the (tailed) Australian cattle dog, became immensely popular. Yet it and its non-pedigree cousins had developed an unsavoury reputation for aggression and being a public nuisance.

Kaleski's real audience were urbanites. They hung on his every word and wanted a cattle dog because they were Kaleski's dog – the rough-and-tumble, macho dingo-dog, the home-grown media star of the day, Australia's first popular muscle dog.

Unfortunately, Kaleski's management advice set a poor example for successive generations of Australians. His cavalier dog-management techniques reflected his literary swagger. He wrote not one sentence regarding conventional cattle-work training, or responsible obedience training and socialisation, or the management necessary for the development of a balanced cattle dog.

As a consequence, too many cattle dogs were owned by too many people who cared for nothing more than the imagined status of owning a tough dog. Nothing handicaps a breed more than fashion-driven popularity.

When not responsibly socialised and managed, the heeler's propensity to protect, chase and heel is manifested in a dangerous form of over-guarding. Over-guarding is self-reinforcing, particularly for a dog that takes protecting its territory so seriously.

Uncontrolled cattle dogs made life hell for postmen, meter-readers, house visitors, passers-by, cyclists and motorists in post-war Australia. Heelers have always figured high in the breeds most responsible for attacks on people, though they are more powerful biters (of legs and hands, usually) than maulers.

Chaining was the common way of restraining working dogs. Containing a dog and restricting its movements is both a good and a bad thing. Good, because it keeps a dog under control and ostensibly safe; bad, because being kept in constant confinement does a dog's head in.

Every dog has its own territorial boundaries and they don't necessarily align with its human-imposed boundaries. Every domestic dog wants more space, and every closely confined dog is vocal in its objections to being restrained. Dogs are social creatures and are never happy alone in confined environments. At times, close solitary confinement is unavoidable, but long term it makes for anxious, hyper-vigilant and needlessly aggressive dogs.

It's the breed of dog, the size of the confined area, the length of confinement, and the management of the confined dog that determine how damaging the experience will be. The smaller the confined area, the more you diminish a dog's ability to naturally protect its territory and itself and to avoid threats.

Chaining is particularly bad for breeds with strong guarding instincts. Whereas working kelpies and border collies appear to tolerate chaining with equanimity, it makes the naturally suspicious cattle dog unnecessarily savage. 'Never go near a chained heeler' has always been the right advice, but it's a sad indictment of the practice. Normal social skills are lost when a dog is constantly chained. When it can't avoid a threat – which eventually becomes anything it sees – and when it can't respond to that threat on its own terms, the chained cattle dog meets all challenges with aggression. The chain-aggressive cattle dog remains aggressive when let off the chain because the damage has already been done.

Occasional, intelligently applied tethering does a well-managed dog no harm whatsoever, and for working dogs and dogs on the back of vehicles, it is absolutely necessary. But permanent chaining has for too long been used by too many individuals as long-term cattle-dog management. In this enlightened age of dog ownership where most people view their relationship with their dogs as a partnership chaining has been generally abandoned.

In the post-war years, heelers were shocking car-chasers. Car-chasing dogs are rarely seen these days, but back in the day the most inveterate car-chasers were blue or red dogs, or mongrels part thereof. In country towns, bored, hungry, uncontrolled heelers often turned to nocturnal sheep killing, many joining the rest of the feral outlaws, with a drastically reduced life expectancy. Municipal pounds were full of them, and far too many met early and untimely ends – euthanised if they made it into the pound, but more poisoned, shot, run over or killed in dog fights.

* * *

The Hall's heeler with the longest working career was the Timmins biter – which eventually became known as the stumpy or the stumpy-tail. There were always stumpies holding down full-time jobs in the bush on large holdings where cattle were still hard to handle.

But for decades the dog-show world virtually ignored the stumpy. Why? Because everyone accepted Kaleski's erroneous claim that two tailed dogs had been the progenitors of Hall's heelers, and so the stumpy was viewed by the ill-informed show hobbyists as a genetic aberration. The bushies who knew their dogs must have been laughing their heads off. For some time no one in the show fraternity seemed to understand that Thomas Hall had produced both tailed and tailless heelers because he put a tailless dog to the tailed dingo. But time would catch up with the myths and uncertainty.

Meanwhile, in the post-war years, the show version of the stumpy increasingly lost ground to its more popular tailed cousin. In Queensland, it was de-registered as a breed in the early 1960s, and it was only the dedication of Mrs Iris Heale of Brisbane that preserved it.

By the early 1980s, the stumpy faced extinction. The ANKC started a redevelopment program in order to save the breed.[3] The program sorted the chaff from the wheat and identified the dogs that were true to type. The stumpy avoided extinction – just.

But the program concluded prematurely and things are still looking grim for the Australian stumpy-tail cattle dog. In 2016, there were only seventy-eight registered nationally. Soon it will only be the bush-bred stumpies that remain.

The stumpy is a wonderful dog, and while a few dedicated breeders and enthusiasts still appreciate it, it is apparently an acquired taste, seemingly not as aesthetically pleasing as its tailed cousin. Temperamentally, though, the two breeds are identical. Had the stumpy come to the attention of the show fraternity earlier, it might well have had a brighter future than it now has.

The great irony is that Kaleski pined for the nineteenth-century-type heeler and mourned its loss, but it was under his nose all along. It's a pity he couldn't see the stumpy for what it was. He would have had the old-fashioned dog he wanted, and he could have done much to promote it. He could have had his cake and eaten it too.

By the time Kaleski was complaining about the lack of respect for cattle dogs as 'the backbone of the beef industry', the beef industry was no longer dependent on them, and the show strains were just suburban pets. Yet other than jostling for the position as the breed's patron and chief authority Kaleski did nothing to ensure the breed's future as a working dog.

The names of great kelpie trial winners could fill a hall of fame, but the cattle-dog fraternity in the bush and in the cities utterly failed to exploit the trial concept to promote the breed's working ability. Selective breeding could have produced a more considered working cattle dog, one that wasn't so savage and stressful for stock.

Cattle-dog trials could have been Kaleski's greatest vehicle for publicising his cattle-dog ideals. He was familiar with sheepdog trials and reported on them, but apparently he did not think to apply the same promotional concept to the cattle dogs. The trial ground would have been the proper forum to demonstrate the worthiness of the breed – not having it pace around a show ring competing for ribbons for the prettiest-looking dog.

The Hall's heeler was a great working breed in its day, but its future as a working dog was finished as soon as the show fraternity got their hands on it. We are left to ponder what the Australian cattle dog could have been if it had had the kelpie's good fortune in the article of its early patronage.

* * *

Kaleski was just as involved in promoting the show version of the kelpie – but he found himself dealing with pragmatic graziers and bushmen, not urban hobbyists, as with the Hall's heelers. The difference between the kelpie and the cattle dog was that the working kelpie men were organised, and continued to support their breed through the sport of sheepdog trials. The working kelpie has proven to be a far more versatile dog than the cattle dog.

In 1902, Kaleski compiled not one, but two breed standards – one for the kelpie and one for his separate 'breed', the Barb. Working kelpies were certainly exhibited in the Sydney show rings before this, and the creation of the kelpie standards, like his cattle-dog standard, would have been initiated through dissatisfaction with judging decisions. Not that breed standards have ever really remedied that problem.

His breed standards officially announced the kelpie, as they had the cattle dog, and defined what a kelpie was supposed to look like, and how it should be judged in the show ring. They gave impetus to the kelpie as a popular exhibit. But, as had happened with the heeler, the demands of the show ring caused a divergence in type from the working kelpie. After a time, show exhibits change in conformation and lose their working ability.

Tony Parsons cites an excerpt from an article titled 'The Kelpie Story', written by Stephen and Mary Bilson of Noonbarra Working Kelpie Stud in Orange, New South Wales, which broadly describes the origins of the show kelpie:

> The original Kelpies, which were working stock dogs, are now referred to as the Australian Working Kelpie and a breed that broke away in the 1920s–1930s is called the Show Kelpie. The Show Kelpies are a specialist line of dogs developed especially for winning dog shows. Obviously, two breeds with the same name 'Kelpie' have caused a lot of confusion among the general public.[4]

Perhaps it is drawing a longish bow to suggest that the working kelpie and the Australian or show kelpie are two different breeds. Anyone viewing either would consider it a kelpie. The truth is that both *are* kelpies, but one strain diverged from excellent working lines to become a show exhibit and companion dog.

Yet the show kelpie can trace its ancestry to some very good working dogs, particularly the great Red Hope, who won a great many sheepdog trials throughout the 1920s.

Supporters of both types point derisive fingers at the other. The working kelpie people claim the show version is useless as a worker and therefore no longer a kelpie. They're half right.

The show fraternity claim that the working kelpie is a crossbred dog. They're half right too.

Tony Parsons is best placed to provide the most balanced view of the division between the two kelpie types:

> I'll briefly be the devil's advocate for the show kelpie fraternity who believe that the show or bench kelpie is 'purer' than the working strain dogs and, if it matters at all, they've got a point. In actual fact 'purity' isn't worth a spit if it isn't accompanied by genuine working ability.
>
> The show kelpie people, or those passionate about their strain, believe that the working kelpie is a crossbred dog, and to a certain extent they're right. Many of the best working kelpies, some of the immortal dogs and always described as kelpie, were in fact part border collie. Johnny, the most famous trial kelpie of the post-war years, was a cross kelpie. Moreover, the Haynes strain of dogs, from which some of the best kelpies in northern New South Wales were derived, had some collie in them.[5]

As we've seen, dingo and border collie were introduced to some kelpie strains; for instance, Jack Goodfellow put dingo into his kelpies. And, Tony Parsons says, there aren't many working kelpies that don't trace to either a Goodfellow kelpie or one bred by a northern New South Wales kelpie breeder. In which case, there's a good chance they've got dingo or border-collie blood, and 'if you outcross your kelpies to a different breed, either border collie or dingo, you change the makeup of the breed, which is what supporters of the show kelpie have said over the years; that is, that the working kelpie is a crossbred dog and the show kelpie is "pure"'.[6]

Mr Parsons has a point, but all modern pure breeds of domestic animals started as cross-breeds. Then at some point, they started to be considered pure, usually when they were breeding consistently to type. Following an out-cross, repeatedly breeding successive generations back to the required breed produces a dog of 93.75 per cent purity after just four generations, although the original type may be permanently altered.[7] A kelpie, say, twenty generations on from one border collie out-cross could hardly be considered to be a crossbred dog. In theory, twenty dog generations could be easily achieved in less than forty years.

So while the show enthusiasts may be technically correct, they are splitting hairs. And what may also be hair-splitting is the working-kelpie mob's claim that show kelpies are not really kelpies because they have lost the ability to work – the very reason for the development of the breed. But is that really true? Show kelpies are still a type of kelpie; they're just not a working kelpie.

It is only to be expected that urban show-kelpie owners who know no better think their dogs are capable of working stock. Yet the show kelpie's build would very quickly prevent it from matching pace and activity with a working kelpie, let alone competing with its working ability. It didn't take long. By the 1920s the show cattle dog's and kelpie's working abilities weren't worth a stamp.

But that didn't stop a lot of breeders of show kelpies from doing irreparable damage to the working kelpie's reputation among buyers from overseas.

Stockmen from the United States who heard about the legendary kelpie and wanted to import working dogs made enquiries with certain show strain kelpie breeders. After reading the kelpie breed standard, they would have been rubbing their hands together. So, show kelpies with no working ability were sent to the US to compete with high-quality working border collies imported from the United Kingdom. As Tony Parsons relates in *The Kelpie*, 'The red show kelpies were treated like a joke.'[8]

A dog that gets excited at the sight of stock and rushes at and around, and through a mob does not a working dog make, no matter what its overjoyed owners think. Working-kelpie breeders produce enough dogs to know that the working ability of the parents is no guarantee that every puppy will exhibit the same temperament or ability (though duds are often made, not born; poor handling of even very young puppies can wreck them). It can hardly be expected that non-working show dogs will produce workers of any ability. But some members of the show fraternity just didn't understand that.

* * *

Despite some observations of dog shows and show-strain working dogs in this narrative it is only fair that readers understand the vital importance of dog shows in maintaining the purity of pedigreed dogs and promoting dog breeds in general.

Without regulated breeding and the ideal of producing *quality* specimens, the standard of many breeds would suffer irreparable damage and some breeds would disappear completely. Yes, some show breeders have been responsible for the dreadful deterioration of the health of some breeds. The ANKC and affiliated state kennel clubs recognise that and are trying to address the issues, but not all breeders are to blame.

Exhibiting dogs was once one of the most popular hobbies in Australia, yet participation is now in decline. There are various reasons for that, and it is a great pity, but State bodies are promoting other activities such as obedience and agility trials, herding, earthdog, gundog tests, dances with dogs, lure coursing, and endurance. These are excellent outlets for dogs and handlers. Still, conformational dog shows are marvellous spectacles where beautiful dogs and excellent handlers show a commitment to their pets that should only be commended. The wonderful array of pedigree breeds available in Australia is due solely to committed breeders importing dogs at great expense and establishing those breeds here. Many of Australia's best dog people are show exhibitors, sporting participants, and pedigree dog breeders.

* * *

Exhibiting kelpies began as a way of showing off fine specimens of Australia's great sheep-working dog. It all went well for a time, but with the influx and popularity of many diverse breeds, particularly in the post-war era, the show kelpie fell from favour, and people lost interest in showing them.

Today the show kelpie – or the Australian kelpie, as he is known in the show world – is certainly a very smart-looking dog. It's a good thing that at least a version of our canine national hero still graces the show rings of Australia. Spectators at the royal shows don't know he's just the recreational model, and being a show dog is not the sort of job that the working kelpie would be satisfied with anyway; but someone's got to do it.

The show kelpie makes a fine companion pet and family dog, whose domain is the suburban backyard rather than the harsh Australian inland. It is every bit as loyal and intelligent as its working cousin. It is just a pity that most of them wouldn't know what a hard day's toil really is.

The kelpie has always been Australia's little mate. It's an easy-going, high-achieving, healthy breed, an endearing but quirky dog that can turn its hand to anything if it has a mind to.

Some, for one reason or another, decided they were not cut out for sheep work and would rather be companions. Enough of them found themselves as pets to form another loose type: the town kelpie. It is usually black and tan, or red and tan, or sometimes a solid colour, but it is often not of the purest bloodlines. Its sweet, happy nature made it the easy choice for many families. Thousands of Australians have fond memories of childhood adventures with their pet kelpie. Loyal and always on the go, it makes an excellent companion and a reasonable sort of watch dog, and a savage kelpie is rare.

There are countless families throughout Australia who have owned them, not always responsibly, and in the past, it has been a very social and very promiscuous dog. Along with the heelers, most common Australian mongrels seemed to be part kelpie.

No kelpie is content to live a life confined to a backyard or curled up by the back door. The kelpie is a dog with a highly developed sense of fun and adventure. It's a natural athlete, and can still have a happy, fulfilled domestic life – if it finds the right people. It's the ideal playmate for the young and not-so-young. It is most suited to active owners who spend a lot of time outdoors. Plenty of kelpies have distinguished themselves as enthusiastic fielders in backyard cricket games, or have been seen riding waves with their owners on paddle boards and

surfboards, towing their people on rollerblades, or leaping off wharves and jetties. The kelpie is the kind of dog who quickly learns when it's time for a walk, or when the school bus is due of an afternoon.

The kelpie always has a lot going on in its head and needs its people to have a lot going on in theirs. It's that big brain that's helped make it the great dog it is.

Good kelpie people know how to keep their pet mentally exercised. They can be killed with kindness, but it takes a bit of doing. As long as they are kept lean, obedient and active with some kind of job, a kelpie will remain healthy, mentally and physically, and live to a good age.

* * *

While there are countless stories of the kelpie's work-related heroics, there are almost as many about its odd, sometimes human-like behaviour. But none is quite as remarkable as the story of the kelpie who lay fighting for his life in a veterinary clinic in Roebourne, Western Australia, on 18 November 1979. He had taken a strychnine bait, but it wasn't his first brush with death. He was the famous Pilbara Wanderer, popularly known as Red Dog.

Red Dog was born in Paraburdoo in 1971 and was first named Tally-Ho, or Tally. He moved with his owner to Dampier when he was eighteen months old. As a youngster Tally would wander all over town, but never took a liking to anyone.

He eventually decided to make a go of things himself, roaming the town from house to house looking for a feed. He eventually took a liking to a man named John Stazzonelli, a bus driver. He got to know the operations of the buses and became a firm favourite with all the drivers.

On the buses, Red had his own seat – normally the seat behind the driver – and any unwary traveller would soon learn that quicksmart. Courier vehicles would also give him a lift, so it was not unnatural that Red Dog adopted the policy of waiting for vehicles to stop and pick him up rather than walking to his destination. He regularly plied his way between Dampier and Karratha, a distance of 21 kilometres, always by way of the buses or cars that stopped for him.

Red adored John Stazzonelli and followed him everywhere. But the association ended tragically when John was killed in a traffic accident.

After John's death, Red left Dampier and began wandering across Western Australia, becoming well known throughout the State. He would hitch rides in trucks, buses and cars, but always find his way back to Dampier in the end.

Red had many brushes with death. He was once found seriously injured, having apparently fallen off a truck or been involved in some kind of vehicle accident. On other occasions he was shot, baited, ripped up by other dogs and infected with heartworm.

Karratha vet Rick Fenny treated him a few times and developed something of a relationship with him:

> When I commenced practice here, Red was brought in several times during the first four months. After every succeeding visit, I was left more and more confused, as he always had a different owner. Eventually he even came to visit the vet's house of his own accord, though I suspect the attraction was my little red bitch.[9]

On 18 November 1979, Red was brought into a surgery in Roebourne showing the typical convulsive seizures of strychnine poisoning. He was put to sleep when it became apparent that he was not going to recover.

The 6 December edition of the *Hammersley News* reported that Red Dog died on 21 November.

* * *

Red Dog was truly remarkable, in that he appeared to choose his peculiar lifestyle, alternating between dependence and independence, or exploitation and self-sufficiency. His 'Littlest Hobo' lifestyle was endearing and interesting, and created a well-deserved legend, but in truth, Red Dog was semi-feral.

But he was not just any feral dog.

As we have seen, a domestic dog is one that is fully dependent on a human who makes all the important decisions for it. A feral dog is one fully independent of human control. Many domestic dogs are forced into thinking for themselves and independence, and there are only two reasons for this: neglect or physical cruelty. It is the need for food, company or safety that causes a dog to venture away from home.

The more a dog makes its own decisions, the more feral it becomes. Once a dog learns it can get along without human control, it will continue to do so unless it is prevented from escaping. It may join or form a feral pack of other independent dogs and never return home.

Red Dog remained a loner, and in a region where free-roaming dog and dingo packs were common, that in itself is extraordinary. He was a very different dog.

Any other dog in his situation would have drifted to the fringes of society and lived as a wary, unapproachable scavenger. Red Dog was much more sophisticated. He possessed the wilfulness, the decision-making ability and the independence that Tony Parsons has identified as key features of the kelpie, and he possessed them in abundance. He was also a very, very quick learner and a problem-solver of the highest ability. If he had been blessed with an owner committed to getting the best out of him, Red Dog could have been anything.

Red Dog certainly wasn't very close to his first owner; the bond was either broken or very weak. He obviously learned early that life was better when he was on his own. He learned to use people for his own advantage, never becoming close to anyone other than John Stazzonelli, but even he couldn't really *control* Red Dog. By then he was just too set in his ways.

His standoffish attitude towards people indicates that he had a high tolerance of people but, other than Stazzonelli, he had no deep affection towards them. Feral dogs don't usually do much tail-wagging and always distrust people. The kindness John showed to Red Dog went some way to redressing the human failing that had created Red's situation.

The basis of dog behaviour is this: a dog only ever does anything in order to gain something, or to escape or avoid something. No matter what reasoning people used to rationalise Red Dog's behaviour, everything he did was for one of those two reasons. He needed food and shelter, and he learned that paying people attention got him both. His ability to exploit people was highly advanced and truly incredible. It is very difficult to explain much of his behaviour based on old personal accounts. Romantic notions always complicate the extraordinary behaviour of such dogs.

Red Dog lived a poor existence, remarkable though it appeared to be. He was an entire (undesexed) dog, and his wounds would have been from fights over bitches in season. He was shot, possibly intentionally run over and ultimately poisoned because some people hate dogs and feral dogs are a real nuisance. Most people cosseted Red Dog, and he probably pushed his luck too far on the occasions when he came to harm.

Red Dog was only eight when he died. He was still in the prime of his life, but as a feral dog, it is remarkable he survived for so long.

But then again, he *was* a kelpie.

CHAPTER 13

The Trials of the German Collie

The German collie is a unique Australian working dog, and a true working collie in every sense. Today it is popularly known as the Australian koolie. It has also been known as the German coulie and the German coolie. All of these names are accepted, but because the dog is being reviewed here in a historical context, we shall mainly refer to it by its original name of German collie. For a while, anyway.

Describing the German collie as a breed might have been possible during the mid-nineteenth century, but today it may be drawing something of a longish bow. They might be more accurately described as a type, or even a closely related group of types, there being much variation among them. That's because their development over the last hundred years has been influenced by several other working breeds.

The German collie is a very striking and capable working dog, but it has never really had the opportunities enjoyed by the kelpie. It enjoys a growing appreciation, and generations of pastoral families have used no other working breed, such is its reputation. It can turn its hand to both sheep and cattle work, and is highly intelligent and trainable.

The German collie has not been immune from the silly breed-origin theories that have been lumped on Australia's other working breeds, but most level heads have always known the German collie's progenitor came from Germany, probably Saxony.

The Altdeutschen Hütehunde, or old German shepherd dogs, were a loose group of stock-working dogs that varied in type. Captain Max von Stephanitz was the driving force behind the development of the German shepherd dog. His book, *The German Shepherd in Word and Picture*[1] attempts to classify all the old German shepherd dog breeds from which the German shepherd had been developed.

Von Stephanitz identifies a variety of German sheepdog, a smooth-coated Schäferhunde – shepherd's dog – from Brunswick in Saxony. It was quite similar to the kelpie in build and had erect ears. Herr Kapitän describes the merle colour - black patches and black marbling imposed on the grey coat - as tiger 'spots'.[2] The big cats were apparently not the von Stephanitz strong suit.

Nevertheless, *The German Shepherd in Word and Picture* is a very interesting and creditable work, though the German dog-technical terms translate very poorly into English, but it is a far better record of the German working dogs of the nineteenth century than anything Great Britain managed for their own dogs. And it shows us a range of German working-sheepdog types and could also contain the kelpie's Rutherford-collie mystery out-cross.

In 1836, when the first settlers arrived to establish the new colony of South Australia, they brought with them Saxony merinos. There were also significant migrations of German farmers to South Australia throughout the nineteenth century, and some of them probably brought German sheepdogs with them. As mentioned earlier the indomitable Eliza Forlonge certainly brought Saxony merinos with her, and she most likely had Saxony sheepdogs with her too, though it is not recorded.

In 1838, there were just 28,000 sheep in South Australia. Yet by 1847 that number had expanded to 784,811, and by 1850 there were 984,190.[3] Such an incredible increase in sheep numbers implies that there were sufficient numbers of sheepdogs to manage them.

So were German collies present in numbers that would adequately service the burgeoning demand across the colony? It's not very likely. Little is known about German collies in Australia, which suggests they were localised and small in number. The breed seems to have been popular mostly with generations of German families, who probably saw them as something of a link to the fatherland.

The merled German sheepdogs that arrived in South Australia were useful shepherd's dogs, but unsurprisingly, as fencing came into use in the 1870s, they were found to be susceptible to the environmental shock of working huge mobs of sheep on big holdings in extreme heat. Farmers followed the only sensible course of action, which was to crossbreed them with local, acclimatised sheepdogs. After a time, out-crossing would have been needed anyway, for better working ability and genetic diversity, because merle-to-merle breeding is not generally recommended. Breeding merle to merle can produce offspring that are termed 'double merles'. These dogs are predominantly white, or display large areas of white around the neck and head and are prone to deafness and blindness.

There was really only one dog to turn to for help when fresh blood was wanted in the German collie gene pool: the

shepherd's dog, or local Scottish working collie, that was also
the precursor of the kelpie. It would have been out-crossings
with this breed that produced the type that became broadly
known as the German collie.

Since the development of the kelpie, every other sheepdog
in Australia – particularly the German collie – has taken a back
seat in terms of popularity and perceived all-round sheepdog
performance. The kelpie would eventually influence and
dominate other sheep-working types in Australia, but because
it did not appear as a breed until around the end of the
nineteenth century it is highly likely the German collie
remained reasonably fixed in type until then. There was
probably some variation within the Scottish working collies,
just as there was in the Schäferhunde, so it is only reasonable to
assume that the early German collies displayed much variation
in type also.

Before long, they spread into western New South Wales
along stock routes, and the waterways of the Darling River
system. In succeeding decades, while never seriously
challenging the kelpie as the top sheepdog, they continued to
be used by certain stockmen.

We have no evidence of the early German collies' working
style. They were obviously no mugs, and having working-
collie blood they could no doubt hold their own in any
company.

* * *

Twelve years after the founding of South Australia, on the
other side of the Pacific Ocean, Mexico reluctantly ceded
California to the United States. In what can only be described

as the most accursed luck for the Mexicans, gold was discovered there later that same year, 1848.

It is known that sheep were shipped to California from Sydney to feed hungry miners on the goldfields. The long-held (American) theory is that the dogs that accompanied this stock contributed to the dog the Americans call the Australian shepherd.

Now, everyone knows the Australian shepherd dog is not Australian, but no one knows *why* it's not Australian. Some theorists have sunk to Kaleski-like depths by claiming the progenitor of the Australian shepherd was the Basque sheepdog from northern Spain that was brought by the Basque shepherds who allegedly tended merino flocks in Australia. Basque shepherds in Australia? No way, José!

The theory goes that the German collies that went on the ships to California contributed to the Australian shepherd. The Australian Koolie Club claims genetic testing has shown that the modern German collie (the koolie) and the Australian shepherd are related.[4] That may be so, but it is highly unlikely that the German collie accompanied those shipments of sheep. It was still a new, minority breed mostly confined to South Australia, and certainly was not common in New South Wales at that time.

And here's the thing: who'd send valued working sheepdogs with sheep destined for a walk to the butcher's hook? There were capable stockmen waiting to receive that stock in California. To suggest they were dependent on Australian sheepdogs to move stock is fanciful.

Yes, the Australian koolie may well be related to the Australian shepherd, a first cousin even, but it is not the Australian shepherd's progenitor. We are all related to our cousin, but our cousin is (hopefully) not our parent.

There is another plausible explanation regarding the ancestry of the Australian shepherd. Before 1848, California was a Mexican province, and before that a Spanish possession. For a time, Spain owned nearly all the western half of what is now the United States – until the United States stole stolen goods when they decided they needed it more.

The Spanish were major wool-growers; after all, Spain is the home of the merino. The dog that traditionally worked the Spanish merino in the sheep-raising province of León was the carea leonés (Leonese shepherd). It is known that in 1765 and 1771, King Charles III of Spain sent flocks of merinos to his cousin, Prince Xavier, the Elector of Saxony. The carea leonés or its progenitor would certainly have been sent to Saxony along with Spanish shepherds to manage the flighty merinos. It is therefore no coincidence that von Stephanitz's tiger spotted shepherd's dog of Saxony and the Spanish carea leonés are basically the same dog.

It is highly likely that the carea leonés, or a derivative, went to California with Spanish sheep farmers just as the tiger spotted sheepdog went to South Australia with German sheep farmers. The carea leonés is most likely the Australian shepherd's true progenitor. That should settle the account of the Basque sheepdog theorists, though there's nothing like a murky breed origin to encourage enthusiasts into hasty, ill-considered theories. Because, as Herr von Stephanitz would have confirmed, the tiger never changes its spots.

Let's hope someone taps the American Kennel Club on the shoulder and further genetic testing absolves Australia's German collie of being involved in the Australian shepherd's paternity. The Americans could then change its name to something a little less embarrassing. Like the American shepherd.

* * *

It is obvious that German collies were finding their way to most wool-growing regions of the country and impressing people everywhere they went. On 19 October 1899, the Hobart *Mercury* reported on a sheepdog trial in which the German collie worked by Mr Barwick of Oatlands received first prize. It was reported to be a very fine specimen of a sheepdog, and secured within two of the maximum number of points.

The German collie won a lot of friends and admirers in Australia, but that never stopped people from scrambling its history. In 1897, Walter Beilby claimed that the German collie was just another name for the blue-merle smooth collie.[5] That is clearly incorrect.

Being a breed that had a small presence in New South Wales, the German collie appears to have escaped the patronage of Robert Kaleski. Some dogs have all the luck.

In an era when Australia was crazy about making show dogs out of working dogs, the German collie also had the good fortune to remain apparently unknown to the show fraternity. It remained what it should have been: a genuine working dog. The pattern had been set by Kaleski's patronage of the cattle dog and the kelpie; as we've seen, both show varieties were rendered useless as workers.

Domestic good fortune might have smiled on the German collie, but world events would soon wipe the smile from Fortune's face, causing her to frown upon the German collie and the German-Australians who had created it.

* * *

Chaos descended upon the world on 28 June 1914, when the Archduke Ferdinand of Austria was assassinated in Sarajevo, Serbia. That murder triggered the First World War, or the Great War, as it was known until mankind contrived to start a second one just 21 years after the first one ended.

Filial obligation and political urging drew Australians into a conflict on the other side of the world that had nothing to do with them. Of a population totalling less than 5 million, 416,809 men, all volunteers, enlisted. It was martial lunacy, and the biggest commitment, per capita, of any nation involved, and it was a cost Australia could ill afford. The country's strongest, fittest sons, many of them young sheep men, were incited to join the great adventure by Australia's war-mongering prime minister, Billy Hughes.

The destruction raged until the Armistice of 11 November 1918. By then, over 16 million people – more than 9 million combatants and 7 million civilians – had perished in the devastating conflict. Sixty thousand Australian men would never see the blooming wattle or the Southern Cross again. One hundred and fifty-six thousand were wounded, permanently debilitated by gas, or taken prisoner. A generation destroyed.

Slaughter on foreign killing fields was a bitter way for Australia to discover its national identity; there was hardly an Australian family not touched by the catastrophe. And there was no German–Australian unaffected either.

German communities had become well established in South Australia. They were highly regarded by other South Australians, who referred to them as 'our Germans'. In country towns, German clubs had become centres of social activity that welcomed everyone.

Yet as the war progressed, the Hughes Government pursued an aggressive internment policy against 'enemy aliens' living in Australia. The majority of those were South Australian Germans. Initially only those born in countries at war with Australia were locked up, but they were soon sharing bunkhouses with naturalised immigrants, Australian-born descendants of immigrants and British subjects born in an enemy nation.[6] To its shame, Australia interned 4500 German-Australians or Australians of German descent during the First World War.[7]

The war also provided Hughes with an opportunity to eradicate German trade influences in Australia. German-owned businesses were appropriated, including pastoral companies.

In 1939, the outbreak of the Second World War triggered a mass fear of invasion by Germany and later Japan. This led to panic that tens of thousands of Australian residents might become saboteurs or spies. Thousands of Australian residents suddenly found themselves identified as potential threats to Australia's national security. The Second World War saw German-Australians interned again.[8]

Eighty-five years after the First World War, Australia's Governor-General, Sir William Deane, delivered an apology to members of the German-Australian community:

> The tragic, and often shameful, discrimination against
> Australians of German origin fostered during the world
> wars had many consequences. No doubt, some of you
> carry the emotional scars of injustice during those times as
> part of your backgrounds or family histories. Let me as
> Governor-General say to all who do how profoundly sorry
> I am that such things happened in our country.[9]

Sincere without a doubt, and better late than never, but Sir William could have said sorry to the German collie too, because internment probably wrecked the purity of the breed.

After the First World War, anti-German sentiment in Australia was high, and wouldn't have helped the German collie's popularity. It was an age of nationalistic mindlessness that may have led to the mindless destruction of many good dogs.

The kelpies suffered a decline in quality because of the Great War too. Many kelpie men never returned from their great overseas 'adventure', and a noticeable loss of the old kelpie types became evident post-war. But the German collie, with its chief support base behind barbed wire in various army camps, would have fared a whole lot worse.

The Great War changed everything for the German collie. With its gene pool drained, out-crossing became necessary again, and the original, old-style German collie would become much diluted after the Great War.

There was no going back to the working collie; the kelpie had totally superseded it. Because it shared the same Scottish working-collie ancestry, the kelpie was a very similar dog to the German collie, if not a close relation. It was the obvious choice to ensure that the German collie remained a superior working dog.

The German collie was also out-crossed with the border collie, and this had a strong influence on the German collie's appearance. The white border-collie tuxedo markings, also known as Irish trim, are a common feature of many of today's koolies.

Because the German collie carried a high proportion of kelpie or border-collie blood, from the 1920s two distinct types emerged.

The kelpie-influenced types are prick-eared, smooth-coated and similar in appearance to a working kelpie. The border collie types are medium- or smooth-coated, sometimes with feathering, and often with white border-collie tuxedo or Irish trim markings and rose ears.

Working-koolie owners never seemed reluctant to out-cross them with other breeds to achieve a dog suited to their particular needs. Some modern koolies exhibit cattle-dog features and are more solidly built. Koolies generally manage cattle quite well.

These days, some koolies show features of kelpies, border collies *and* cattle dogs. Merles and solid colours prevail in all types, with blue merle still being the standout colour. The merle gene affects eye colour. Koolies can have one or two blue eyes, or sometimes green or yellow eyes with blue chips.

The variation in koolie type can make it difficult for even a trained eye to recognise some individuals as being of the same breed. It is this variation that challenges the strict definition of a breed – a group of dogs that all have similar conformation and reliably breed true to type.

The German collie is also sometimes confused with the kelpie, since they often share the same solid colours and a similar build. The German collie's striking blue or less common red merled coat is its most distinctive feature, but solid-coloured dogs are vitally important for the health of the breed. There have always been solid-coloured German collies, though infusions of kelpie blood may have reinforced solid colours among the population.

There is an old bushman's saying: a good dog cannot have a bad colour. Unfortunately, in common with other merle-coloured breeds, koolies can, under some circumstances, have a bad colour. The Koolie Club of Australia has gone to lengths to ensure that potential koolie buyers or breeders fully understand the genetic issues related to merle-coloured dogs, and provide access to budget-priced genetic testing for members' dogs. It is an excellent initiative because ill-considered breeding may be the koolie's biggest threat.

The Koolie Club has no plans to seek ANKC recognition for the breed. Preserving the koolie's working ability is its chief objective, and one that should be applauded. But the koolie does have a professional, broadly descriptive breed standard. It is quite similar to the kelpie breed standard, in that it describes a conformation and temperament ideal for a hard-working Australian stock dog. Here is a general description kindly provided by the club:

> The koolie is first and foremost a functional canine working breed. The ability to perform its purpose is wholly dependent on the combination of the key features evident in the breed of today. The fact that the koolie has the combination of features necessary for performing top quality herding work is no accident. Breeders over the generations have been very selective in choosing breeding stock on the basis of field performance.
>
> The general appearance should be that of a strong, active, and athletic dog, well-muscled and in hard condition, combined with great suppleness and agility, indicating the likelihood of stamina and the capability of untiring work. Any coarseness or weediness is undesirable. Given the range of work of the koolie, e.g. from herding sheep or cattle in the open field to yard work and trucking, there may be slight differences in desirable

conformation and various family lines may differ slightly accordingly.[10]

But here's the koolie conundrum. Standards are good things in theory and bad things in practice. There is so much variation that koolies differ markedly, not slightly, and a lot of current specimens may not meet the standard.

The out-crossing practices so common among koolie owners are the biggest impediment to establishing even a broad interpretation of a single basic type. For a purely working dog, though, that is no big deal.

* * *

The *Weekly Times* of Melbourne provided some interesting information about the German collie in its 2 May 1951 edition. A reader had requested information about the breed, which prompted another reader, Mrs M Buesst of Toorak, to offer some details of a dog she had once owned that she believed to have been a German collie. Mrs Buesst said her dog had been named Onyx because of his colouring. She said his coat had featured patches of black and white, which gave him a blue look. Onyx had been about the size of a Queensland heeler (a cattle dog), but more graceful, with one eye blue and the other brown.

At first she had thought Onyx was a mongrel heeler, though he had not heeled anything. Old men who saw Onyx insisted he was a German collie, though Mrs Buesst had never heard of the breed. Mrs Buesst was then told that German collies were brought here in the early days to work sheep, but had proven timid and hard to train, and had gone out of fashion. Rough treatment or a harsh word was fatal with them, according to her ill-informed informant. Yet Mrs Buesst said Onyx was the most intelligent and best dog she ever owned.

Mrs Buesst's informant only got part of his story right. Koolie men and women have never found the dogs to be fragile. On the contrary, they are as robust and trainable as any of Australia's great working dogs.

The Melbourne *Weekly Times* of 30 May 1951 gave a German-collie breeder a free plug. Mr E E Small of Allambie Station in Gippsland wrote to say he had three German collie pups for sale. Mr Small said of his dogs, 'I have found the breed reliable, clever and intelligent. They make splendid watch dogs. My pure bred, which I bought in 1949, is a perfect sheep dog especially with ewes and lambs. True colour for the breed is a lovely, slate-grey blue, black spots and wall [white-coloured] or brown eyes. Bitches may be whitey grey with coloured eyes.'

* * *

It is not known when the German collie started being referred to as a 'coulie' or 'coolie', but they were still commonly called German collies by some people as late as the 1950s.

There are two theories on the origins of the coolie name. The first is that it was the German–English rendering of 'collie'. But 'coolie' is also a derogatory term for an Asian, usually Chinese labourer, so the second theory goes that the dogs were named after them because they worked like coolies. No one really knows, and the breed club decided to go with 'Australian koolie' to avoid confusion with collies.

The Australian koolie, though, is very much a genuine Australian working collie. There is nothing it cannot be taught; it is limited only by its owner's ability to bring it to its full potential. It is employed all around Australia in a variety of stock-working jobs, and has also started to attract interest from the United States. The koolie's future in the Australian pastoral industry is assured, even though it faces stiff competition from the more widely used kelpie and border collie.

In common with Australia's other great working dogs, it is a natural athlete built for endurance, though it is thought by some to be generally less intense than the kelpie and the border collie. Some take time to grow into a working dog, and some seem more suited to the role of faithful companion. Koolies benefit most from active, thoughtful owners who understand their need for a steady, regular job. They are strikingly attractive dogs that develop a loyal following among the people who own them.

The Australian koolie has a rich, unique history, and deserves its place as one of the dogs that made Australia.

But for every British, German or Australian collie that toiled to build a new nation, there was another canine – no working dog – that did all it could to destroy the new order. It was the dingo, and it brought real war to Australian pastoralism, unlike the mythical terrors of the bush that sprang forth from Robert Kaleski's imagination-powered typewriter.

CHAPTER 14

The Wild Dog Wars

Scientists believe that dingoes are one of the most important components of a healthy Australian environment. It is a phenomenon known as the 'trophic cascade' – the principle that top predators change the behaviour and numbers of herbivores, which then affect the vegetation and the soil composition of a habitat. It cannot be denied that Australia's most balanced environments are those with a sustainable population of dingoes.

But you wouldn't find a wool-grower in Australia who would agree – and you can hardly blame them. Australia has always been in love with wool, and the dingo is wool's Big Bad Wolf. Yet the environmental degradation of the sheep lands of Australia's interior is believed to be directly linked, at least in part, to the absence of dingoes.

Colonialism and pastoralism demolished the natural order and painted the dingo into a corner. Yet the new order's soft underbelly was only really exposed when pastoralists tried to tame the harsh interior and grow roots in the parched land of the nomad. Wool should have been off and running, but this was Australia, and the dingo proved to be the most wearing of a myriad of complications.

In terms of numbers, the invaders were an irresistible force, but the isolated wool pioneer had all the power of a besieged king on a chessboard, with a range of hostile pieces arrayed against him. And there was no piece more hostile or mobile than the dingo. It extracted an excessive, human-like revenge on the pastoralists, who, despite crushing the established order, were knocked from pillar to fence post.

Even if the dingo had killed only what it needed to survive it would not have been tolerated. Yet the mass killings continued and it was evident that the dingo would not be defeated by the gun and trap alone. For a time, the dingo held sway and things were touch-and-go.

The emotional effect on a grazier confronted with the continual killing and maiming of his or her stock should not be underestimated. A 2013 study reported that:

> The Victorian farmers' stories had common themes of frustration, loss, grief, a sense of powerlessness, lack of control and helplessness which were indicators of the psychological pressure these farmers attribute to wild dog attacks. There was also increasing pressure to fence vast areas, spend more time staying out at night to protect stock, more time talking about the issue, and more time recording statistics and telephoning wild dog controllers to report attacks. The majority of those interviewed indicated that it was not so much the financial loss that affected them, but the anxiety and stress of finding sheep ripped and bleeding. Others spoke of being hyper-vigilant and the effects of losing sleep which can result in poor attention, concentration and memory, irritability and other mood disturbances, and impaired judgement and reaction time. For some, the anxiety felt had escalated to fear for personal safety.[1]

Stock attacks would have taken an even bigger toll 150 years ago, when graziers suffered far greater isolation in an attempt to make a new life for themselves and their families.

Eventually, the problem in colonial Australia became so acute that governments had to step in to support the wool-growers. The battles raged across the breadth of the continent and, in some cases, wool's viability hinged on the eradication of the dingo.

* * *

It was a virulent alkaloid that tipped the apothecary's balance in favour of the settler. Strychnine poison became the grazier's lethal ally. Available in crystalline form in Sydney for 30 shillings an ounce (equivalent today to about $240 for 30 grams), strychnine saved graziers hundreds of pounds a year in stock losses, not to mention relief from the gut-wrenching grief of seeing their futures being nightly torn to pieces.

The popular method of baiting dingoes involved dragging a dead sheep, kangaroo or emu behind a wagon or vehicle at sundown, and leaving baits at intervals of a mile or thereabouts. The baits were pieces of fresh meat, onto which a small quantity of the poison had been sprinkled. The meat was tied to an overhanging branch by a thin piece of twine, to keep it out of reach of the feral cats (which had been found helpful in destroying small native animals). The suspended baits were set so that the dingo had to get up on its hind legs to take them.

The dingoes would follow the carrion trail and be encouraged by small pieces of unpoisoned meat dropped occasionally along the path. At the end of the trail, strychnine was sprinkled over the organs of the carcass.

It was a very effective way of thinning out the dingo population, particularly if a sheep carcass was used as the drag bait. It killed a lot of other vertebrates too, the wedge-tailed eagle chief among them.

In New South Wales the slaughter went wholesale, and strychnine became a weapon of mass destruction. Killing one adult dingo, particularly a nursing bitch, indirectly killed several more juveniles, but it is probable that baiting mostly thinned out the least wary specimens – probably young, inexperienced or aged individuals looking for an easy feed. Hard-wired, sheep-focused killers may have ignored the baits and boring old carrion when there were live sheep to kill.

Strychnine was used widely for at least a century to control dingoes, but it also found another use: decimating the large populations of feral dogs in and around Sydney and other cities and towns. Strychnine was cheap and easy to obtain, and residents incommoded by their neighbours' uncontrolled domestic dogs often took matters into their own hands and used it to destroy those dogs as well.

Losing pets to baiting was a common occurrence while strychnine was easily obtainable. The local 'dog baiter' was a common, usually anonymous, and feared presence in many suburbs and towns. It was a cruel and unethical practice (the end not justifying the means) but it largely eliminated the feral and uncontrolled dog problem – for a while. And it caused many 'she'll-be-right' dog owners to reconsider the wisdom of letting their pets roam at large.

Then in 1968, 1080 (ten-eighty), an environmentally sensitive and target-specific toxin, became the favoured agent for graziers. It is an odourless compound that occurs in approximately thirty species of native Australian plants.

1080 is still widely used for wild dog control, and while it is said to be particularly effective, dogs and dingoes can be extremely difficult to locate once they have taken a bait as it can take up to three hours before the poison takes effect.

* * *

When it comes to ridding sheep lands of dingoes, the reality is that only two methods work. The first is extermination, which is easier said than done, though the rural sector has never stopped trying. The second is exclusion.

Around the time the kelpie emerged as a breed, the dingo had been largely exterminated from most of inland New South Wales, Victoria and parts of southern Queensland. At the same time, the northward expansion of the rabbit plague was overrunning sheep country. More dingoes might have gone some way towards slowing their progress. Oh, well.

In desperation, the rural authorities in Queensland decided to construct a rabbit-proof fence across the southern third of the State in 1884. It didn't stop the rabbits, but it kept out wild pigs, big marsupials and emus. Having pushed the dingo mostly north of the rabbit fence, graziers to the south lobbied for the fence to be heightened to also exclude dingoes. It didn't seem like such a bad idea – for the graziers at least.

The Dingo Barrier Fence, as it became known, started at Jimbour in the Darling Downs of southeast Queensland. It zig-zagged its way in a northwesterly direction to the east of Mount Isa before turning southwest and following property boundaries to the New South Wales border. Running west along the New South Wales–Queensland border to Cameron Corner, it continued south down the New South Wales–South

Australian border before turning west and zig-zagging across South Australia, trending south, to end in the Nullarbor Desert at the very cliff-edge of the Great Australian Bight. South Australia added the 2150-kilometre-long section of the fence to protect the southern sheep lands in the late 1940s.

The Dingo Barrier Fence is 180 centimetres high and 5614 kilometres long – the longest construction in the world. It allowed southern and central Queensland, the majority of New South Wales and Victoria, and southern South Australia to rid their sheep lands of the dingo.

For a while, everything seemed rosy. But the Dingo Barrier Fence was just a Band-Aid solution. Dingo-proof fencing is ridiculously expensive to build and maintain, and there is no guarantee that dingoes will be entirely absent from the protected areas. Holdings with the barrier fence as a boundary were still subject to constant infiltration and attack, and small pockets of dingoes survived in rough, inaccessible country inside the protected areas. (In fact, dingoes have never been eradicated from the Great Dividing Range or the adjacent tablelands.)

Peter Waite was a very experienced pastoralist who owned several large properties in dingo-infested, semi-arid South Australia and New South Wales from the late 1860s. He was a frequent contributor to *The Pastoral Review* from the late 1890s to the 1920s. He had this to say about dingo barrier fences:

> No fence will kill dogs. It can only be a barrier, and call a
> halt for the dogs, which gives an opportunity to kill them
> by means of traps and poison. If advantage is not constantly
> taken to systematically employ these methods of
> destruction, it is only a question of time when the dogs

will get inside either through washaways, rabbit holes,
blowouts under the fence, open gates, breakdowns, or
drifted sand. Without constant attention, no fence is proof
against live dogs.[2]

Waite was right, of course. The fence was just the start. Constant vigilance and continuous poisoning, trapping and shooting were needed to keep dingo numbers down and induce the animals to avoid the fence. Maintenance of the fence was left to the graziers, and before long it degraded and was leaking dingoes. There's nothing the dingo likes more than a challenge. A fence wasn't going to keep it contained for long.

By the mid-twentieth century the sheep lands west of the Great Divide were supposed to be dingo-free. It wasn't necessarily so. A dingo known as the Billa Billa Killer went on a sheep-killing rampage for several months in the Goondiwindi area on the Queensland–New South Wales border, and on one property alone caused the loss of 200 sheep. It carried a bounty of £50 on its head, and was trapped and destroyed in March 1939. When caught in a leg trap on Ellenvale Station it was so desperate to escape that it had attempted to chew off its trapped leg.

There were gradual, insidious encroachments, but at least the mass sheep slaughter had abated. With the dingoes mostly gone, kangaroos and emus proliferated. Then there were the rabbits, and they really did denude the landscape until the release of myxomatosis thinned them out in the 1950s, but by then the damage was done.

In the thick red country of the western regions, the next plague was the feral goat. The feral pig mobs also burgeoned with no predators to keep them in check. Pigs are enterprising omnivores, and they killed sheep as well as polluting water, creating death-trap bogs for sheep, and destroying water infrastructure and fencing.

Dingoes regularly preyed on foxes. Without the dingo around, fox numbers too exploded, and the creatures played merry hell with weak sheep and lambs, particularly at lambing time.

Graziers developed a deadly set against nearly all local wildlife, which they saw as some kind of threat. It wasn't that they were paranoid, it was just that everything was out to get them. Wedge-tailed eagles (the graziers called them 'eagle-hawks') copped the blame for carrying away lambs, and the more hysterical raptorphobes warned that these birds wouldn't scruple to swoop down and cart off an unguarded baby too. Shot eagles were strung up on fences (the grazier's raptor version of the human head on the pole) as some kind of mindless warning to other eagles of the consequences of eyeing off lambs or babies.

Then there were crows picking the eyes out of sheep bogged in the mud of waterways and earthen dams, and of course the biggest sheep-killer of the lot: the fly. Bad, mean and, as we know only too well, mighty unclean. Afraid of no one, they got the big leg-up by means of the countless daggy sheep and putrefying carcasses of all descriptions that littered the country. The dirty little thing.

* * *

Without question, the real front line of the dingo war was central Queensland, north of the Dingo Barrier Fence.

The first legislation to regulate vertebrate pests in Queensland, the horrendous *Marsupials Destruction Act*, dates from 1855. In an 1885 amendment, dingoes and certain other native and introduced animals were declared pests. Bounties were paid for the introduced European hare, the European red fox and the feral pig, good, but just to ensure the baby (that

In the Snowy Mountains of New South Wales, graziers and professional doggers hang the carcasses of shot, trapped or poisoned dingoes and feral dogs from roadside trees. (Courtesy of Vicki Hull)

wasn't carted off) went out with the bathwater, there were also bounties on kangaroos and wallabies, pademelons, rat kangaroos, bandicoots, koalas, and of course the notorious lamb- and baby-snatcher, the wedge-tailed eagle. Half a million koalas were destroyed in Queensland alone in 1927. 500,000. In one year. But let us not forget the villain of the piece. Bounties were paid on presentation of a dingo scalp.

The Queensland government conducted extensive aerial baiting campaigns in the twentieth century, dropping approximately one million strychnine baits every year in the pastoral regions. They also facilitated ground baiting by graziers. Millions of native and introduced animals were destroyed over the years.

Gradually government-subsidised bounties were dropped, until only the $10 dingo-scalp bounty remained. Between 1885 and 2001, over 1.5 million dingo scalps were presented in Queensland for bounty payment. The number presented each year varied hugely, from almost 50,000 scalps in 1957–1958 to 2689 in 1975–1976. As those were State-wide figures it is hard to pinpoint the cause for the dramatic decrease in paid bounties. But a drop in the dingo population on both sides of the fence after decades of relentless war is the most likely reason, and more poisoning rather than trapping and shooting could be another.

To obtain the bounty, trapping and shooting rather than poisons were preferred, because poisoned animals are seldom discovered fresh. A study in 1987 found that 87 per cent of the 18,614 dingo scalps presented for bounties were either trapped (50 per cent) or shot (37 per cent).[3] Trapping is particularly labour-intensive yet suited to dingo control in sheep areas, because it makes these areas safer for working dogs. Sheep

producers rely heavily on their dogs to muster and handle sheep, and poison baits create unacceptable risks. Many great working dogs have been lost to wild dog baits.

* * *

The Dingo Barrier Fence in Queensland is now only a fraction of its original extent. A decline in the State's sheep numbers during the 1960s and 1970s and a corresponding expansion in cattle production, coupled with economic recession, led to degradation of sections of the fence. The cost of maintaining a 2-metre-high, wire-netting boundary fence was prohibitive.

The Queensland government decided to upgrade and realign the Dingo Barrier Fence in the early 1980s.[4] It reduced the State's protected sheep-production areas to about half. The livelihoods of wool-growers north of the realigned Dingo Barrier Fence were thrown to the wolves. It was the beginning of the end for wool in northern Queensland.

North of the current alignment, the dingo has won, and the Queensland wool industry has been almost driven to extinction. Devastated wool-growers who lost thousands of sheep a year in the unprotected areas have either converted to beef or grain farming, or left agriculture altogether. Wool is no longer viable in unprotected areas of Queensland.

The dingo is more hated in Queensland than the crocodile, the box jellyfish and the New South Wales Rugby League team. And that's saying something.

* * *

Things are almost as precarious in the dingo's other great stronghold, the Great Dividing Range of New South Wales and Victoria, particularly where wool-growing concerns border national parks. The Northern Tablelands and the Brindabella, Snowy Mountains and Monaro regions of southern New South Wales are under constant assault. Pasture protection boards were involved in the construction and maintenance of dingo barrier fences in New South Wales – albeit small, localised attempts at protecting sheep. The same problems that plagued the big fence in Queensland – a lack of unity, prohibitive expense, and poor maintenance – were evident in the New South Wales region of New England.

The *Armidale Express and New England General Advertiser* of 5 February 1943 reported that graziers around Guyra were extremely concerned at the number of sheep being killed in the district by dingoes. Bounties as high as £70 a scalp had been offered to hunters, but the dingoes were still killing sheep. The killings had caused many graziers to give up wool-growing.

Eastern Australia was affected by drought in 1952. It was a very bad year for dingoes along the Great Dividing Range. On 18 April 1952, the *Newcastle Morning Herald and Miners' Advocate* reported that dingoes were increasing in number and had been attacking stock affected by the drought. A committee for the destruction of the dingoes had been formed in Eccleston district north of the Hunter Valley. The committee had asked the Dingo Destruction Board for two dozen dog traps and strychnine.

The *Glen Innes Examiner* ran a story in May of that year, informing readers that the Moona Pastoral Company, east of Walcha, had had to reduce sheep on its holding from 4000 to

300 and run cattle instead, because of the ravages of the dingo. The pastoral company had been unable to secure netting to fence out the dingoes.

The wild dog wars in sheep country along the Great Dividing Range are ongoing. The problems are so acute in the Monaro and Snowy Mountains regions of New South Wales that lots of graziers have gone into cattle in an effort to minimise their losses. Aerial baiting with 1080 is the most cost-effective way of controlling wild dogs in inaccessible terrain, usually national parks, that borders much of the pastoral land. There is no end to that conflict in sight until either the stock and their owners or the wild dogs are eliminated. And that ain't gonna happen.

* * *

Western Australia also constructed barrier fences to halt the progress of rabbits, which were later upgraded to exclude emus and dingoes.

There are three fences in Western Australia in total. All were constructed between 1902 and 1907 and span a combined length of 3256 kilometres. The first, previously known as the Rabbit Proof Fence, divides the State from north to south, ostensibly providing protection for the arable region of the State's west. The second parallels the main fence to the west and along the lower half of the State. The third, the shortest of the three, runs from the second fence to the coast near Kalbarri.

Like all the other barrier fences built in Australia, the Western Australian fences provided limited protection for the sheep-raising districts, and since the 1980s they have fallen into disrepair.

* * *

More recently, the dingo has earned another kind of bad reputation.

Left to its own devices in his natural environment, the dingo poses no threat to people. It is a shy, retiring creature that happily coexisted with Australia's first people for 5000 years. Even after the British colonisation of Australia, when the dingo was subject to the old-world biases and medieval myths, it was never considered to be any sort of threat to humans.

Old-world wolves, too, avoid people at all costs. They have rarely, if ever, preyed on humans. It is believed that the stories of wolves who have attacked people were instances of rabid animals.

That's not to say the dingo never made a pest of itself when opportunity beckoned, or that dingoes could not show aggression towards people if pushed to it. But as the decades of the twentieth century rolled on, things changed between man and the dingo. Urban Australians and international tourists began to visit remote Australia in large numbers. Lots of them went to those places to see dingoes, and none of them was ever warned off feeding them.

That inappropriate interaction would cause people and dingoes to lose their mutual wariness. Familiarisation breeds contempt, and there's no more contemptuous animal than a spoiled dingo. The camp-ground dingoes would begin to pose a genuine threat to humans.

Human-directed dingo aggression is always caused, directly or indirectly, by people. People never fail to find innovative ways of making animals aggressive towards them. Inappropriate habituation – the process by which a wild animal is conditioned to lose its fear of people through regular, inappropriate contact – is the sole cause of recorded attacks by wild dingoes on people. Inappropriately habituated wild dingoes treat obliging humans with derision. That derision is manifested in aggressive behaviour when the canid senses an opportunity to exploit weakness, and it's usually about food.

Without exception, dingo attacks at tourist venues have occurred by dingoes that have been fed by tourists. Feeding wild animals has caused serious safety problems at tourist venues all over the world. It is a stupid and dangerous practice because it creates a dependence on people for food, and teaches dingoes that if they push the point, people will submit to them.

As we've seen, adult dingoes were never tolerated around camps by Aboriginal people. Pet dingoes were evicted when they reached sexual maturity if they didn't take the hint and leave of their own accord. Wild dingoes have no place in close proximity to people. They are opportunistic, bold and highly intelligent, and only ever hang around people to exploit them. And dingoes are unique among the wild canines, because they have had 5000 years to adjust to exploiting people.

Dingoes that live in close proximity to tourist venues are taught by their parents to make a living this way. A dingo in a camp ground is hunting and views all the food resources within its territory (which includes the tourist park) as its own. These dingoes may have lost other hunting skills and might rely solely on people for their existence.

The dingo's appearance and behaviour disarm naïve tourists. There are few wild canids that will confidently approach people like a habituated dingo does. It is very dog-like behaviour, but the dingo is no dog and has no affection for people. And because dingoes in the Red Centre are usually thin and sometimes mangy, tourists used to stuffing food into their pet dogs feel sorry for them. It's an anthropomorphic recipe for monsters in the making.

On the night of 17 August 1980, a dingo took baby Azaria Chamberlain from her family's tent in a camp ground in Uluru–Kata Tjuta National Park in the Northern Territory. Baby Azaria's body was never found, and on 29 October 1982 her mother Lindy was convicted of her daughter's murder and sentenced to life imprisonment. Her husband, Michael, was given an eighteen-month suspended sentence for being an accessory. When Azaria's clothing was found in a dingo's lair in early 1986, a Royal Commission was established to review the case. On 2 June 1987, Justice Trevor Morling cleared the Chamberlains of any culpability. On 15 September 1988, the Darwin Supreme Court quashed all convictions against the Chamberlains.

Fraser Island in Queensland also has a real problem with habituated dingoes and tourists. In April 2001, two dingoes killed a young boy and mauled his brother. This led local rangers to cull thirty-one dingoes and implement a range of strategies to keep tourists protected. There have been a multitude of other threatening incidents reported concerning dingoes on Fraser Island.

In another instance, a thirteen-year-old girl was attacked at night while camping at a Northern Territory caravan park.

Unfortunately it's true that habituated dingoes pose a real threat to children. As a predator, the dingo is instinctively attuned to identifying weaknesses in prey. It is now apparent that habituated dingoes view weak people, particularly children, in this way.

Inside a national park, where dingoes can become habituated and dangerous, they are protected. Outside of national parks, where the dingo poses no threat to personal safety, it is a declared pest and can be shot on sight. It is a very sad indictment of the management of Australia's apex predator.

So how much of a risk to humans is the dingo compared with the domestic dog? A September 2005 study of dog-related injuries conducted by Renate Kreisfeld and James Harrison of Flinders University in Adelaide[5] found that death resulting from dog-related injury is a rare event. During the seven-year period from 1997 to 2003 (inclusive), eleven deaths caused by dogs were registered. The majority of these cases involved male dogs. The deaths were fairly evenly distributed between young children (zero to nine years), older children and adults (ten to seventy-four years) and the elderly (seventy-five years and over).

These eleven fatal attacks all involved domestic pets. That compares with only one fatality caused by a habituated dingo over the study period, in 2001.

The dingo is a creature of the bush, and to dominate its environment it must be suspicious and serious. The dingo will never accept domestication because it is a wild creature – a wolf. The problem is people want to make a dog of it because it looks like a dog. It can be agreeable enough when sensitively kept, because it knows on what side its bread is buttered, but it has none of the dog's devoted servility. It will always prefer its own kind to humans, and the call of the wild to its master's voice.

But it now seems that if the dingo is to survive in its pure form, it *must* be kept by humans. Soon captivity may be the only place to see a pure dingo.

* * *

Dingoes, as we know, readily mate with domestic dogs. As mentioned, however, the domestic dog cannot establish a viable wild population without hybridising with the dingo. It's the reason why feral dogs have never been able to establish themselves in Tasmania, where dingoes are absent.

The feral dog, the dingo's partner in crime, is its real enemy. Hybridisation spells the end for the genetically pure dingo. Virtually every dingo today is a hybrid mongrel, except for those in managed reserves or in the most remote parts of Australia, but even their purity may be compromised by camp-dog blood.

As we have seen, when mated with the dingo, the domestic dog's appearance succumbs to the dingo's powerful genetic influence. Colour appears to be the only feature that will possibly change in the outward appearance of the early hybrids. But while a hybrid may maintain their dingo's appearance, it has destroyed the dingo's genetic purity.

The dingo was once called the wild dog, but these days 'wild dog' is a generic term that can refer to the dingo, the wild or feral domestic dog, or the dog and dingo's hybrid progeny. Australia is experiencing a wild-dog crisis, and more than just the wool industry is under threat.

The increasing feral dog problem is human generated and fuelled. There are several reasons behind it. The first is incompetent management. The second is the popularity of pigging dogs. The third is the explosion and expansion of feral pig populations – and that environmental problem can be partly attributed to the removal of the dingo from most of southeast Australia.

It seems like every second four-wheel drive utility vehicle in the bush has a dog cage on the tray. Pigging with dogs is an extremely popular pastime these days. There are several magazines devoted solely to the pursuit.

Anyone can own a four-wheel drive and have a cage attached to the tray. And anyone can own a pigging dog – but not everyone can properly raise, socialise, train and manage one.

There is a mind-boggling array of homegrown pigging 'breeds' available in Australia today. Different pig hunters and wannabes have their own ideas of what breeding makes the right type of pig dog and many sorts of large hunting breeds are used. They are always muscular, athletic types, though the muscle between their ears seems to be the weakest in some of them. The same goes for some of their owners. The most popular type is the bull arab, a dog that has nothing to do with Arabs. They were developed in the 1970s by a Queensland man named Mike Hodgens, and the originals comprised half English bull terrier, quarter greyhound, and quarter German shorthaired pointer, though plenty of variants display boxer, English pointer, Irish wolfhound and Scottish deerhound blood – whatever takes your fancy. They are all excellent feral pig wranglers, but the bull arab and other similar pigging types are also dogs whose reputation has been tarnished by some of the fools who have owned and bred them.

There are obviously a lot of great dog owners in the pigging community. And there are some communities and regions where the piggers take their pig-dog owning and pigging responsibly and seriously. The New South Wales New England town of Armidale is one such place. There the local piggers maintain their own registry of pigging dogs and tend to mostly self-regulate in cooperation with the local council.

But no matter where feral pigs are found there are a hell of a lot of dog mugs mixed in with the good operators.

Untrained, useless pigging dogs regularly become lost in the bush or are dumped by their irresponsible owners because they are not hard enough. Even well-trained dogs can become well separated from their handlers in the heat of the chase, so switched-on operators have GPS tracking fitted to their dogs.

Some lost pig dogs do make it out of the bush. They often turn up emaciated, wearing hunting collars and chest plates. They constitute a high proportion of the urban feral population, and country pounds are full of them. Rehoming them can be problematic.

Two of the most common coat colours among the pigging breeds are white parti-colour (coloured patches on a white coat) and brindle. These are also the predominant colours appearing in dingo hybrids. One does not need to be a rocket surgeon to work out that lost pig dogs make a significant contribution to the feral-dog population in areas where feral pigs are prevalent. The introduction of pigging-dog blood has added a whole lot of muscle to the wild-dog problem.

One worrying influence of such muscle-dog blood on the dingo gene pool is the lessening of the dingo's natural fear and wariness of humans. The hybrid wild dogs seem to be bolder and more fearless than the dingo would be, left to its own devices.

The National Wild Dog Action Plan has been developed to address and coordinate the general wild-dog problem on the Australian mainland.[6] Annual stock losses across Australia are appalling, and increasing every year. It is estimated that in 2009, losses to the cattle industry were $26.7 million, and $21.9 million to the sheep industry.

The wild-dog threat has now broadened to semi-rural and urban areas, with pets, stock and people being threatened and attacked. It's a worrying development. Today, wild dogs are found throughout the Australian mainland, and are causing problems in regions that haven't seen dingoes for over 100 years.

Something certainly needs to be done, and unified action is the way to go. But in the process, another baby may well get thrown out with the bathwater.

* * *

The wild dog has waged guerrilla war against pastoralists for over 200 years, and that war shows no sign of ending any time soon. It demonstrates that medium-sized animal farming in Australia is impossible without taking war to the environment. The whole thing is an unfixable mess, because one side provides all the rations the other side needs.

Wool might have made modern Australia, but wool and the wild dog have paid a heavy price for the sake of national prosperity. The losses to both sides have been appalling, with millions of sheep and wild dogs slaughtered. Yet it seems it is the grazier who can least bear attrition's cost. Australia's Big Bad Wolf and its feral cronies have huffed and puffed and blown down those homesteads made of straw and sticks.

If some pastoralists had their way, the dingo would be driven to extinction, and its melancholy howling would never again be heard in the Australian night. The graziers would be happy – but not for long, because their sheep wouldn't be any safer. There would still be feral dogs, as well as foxes and flies. The environment would further degrade to irreparable imbalance, and then, when the Australian climate and El Niño felt like it, they would conspire to send drought or fire or flood to decimate the mobs.

The dingo and the pastoralist are both invaders. One tiptoed into the bush. It fell in with the beat of the environment's drum and conquered the continent. Nothing changed but the predatory hierarchy. The other marched in to the beat of its own drum and defied the environment, and it has been bogged down in a war against that environment ever since. Australian pastoralism has paid a heavy price for its success.

It was Australia's great dog authority Robert Kaleski who created the relentless wild dog smear campaigns that fooled the nation. And like a seasoned wartime propagandist he sought to turn the graziers' 'enemies' – one real, the other imagined – into monsters.

CHAPTER 15

Bush Myths and Monsters

While the dingo wars raged on, urban Australians had no idea what was really happening in the bush. They had a completely different attitude from their compatriots losing thousands of sheep a year to dingoes. They were fascinated with the dingo wars, and dependent on the media to learn more.

There were attempts by the Sydney press to provide readers' questions regarding Australia's native fauna. Arthur Crocker, a popular author of the time, attempted to do just that in Sydney paper *The World's News,* on 15 January 1916.[1] A reader identified as 'FEH', of Neutral Bay, wanted to know how the dingo came to Australia. Mr Crocker quoted the experts.

Ernest Le Souef, the Director of the Sydney Zoo, had a crack at answering. The dingo is not a native Australian, he said, but an Asiatic dog that came in from New Guinea at a time when Australia was joined to New Guinea, but after Tasmania was cut off by the sea. It was not such a bad response. Mr Le Souef got the dingo's origins right, but pre-dated its arrival in Australia by many thousands of years.

Decorated First World War soldier Captain Walter Harris was an active bushman, and before the war he had authored a book, *Outback in Australia.*[2] In the book, Captain Harris was pretty firm in his opinion on the arrival of the dingo. He said it was the descendant of a few domesticated Dutch dogs that were left in a diseased state on the shores of Western Australia by the officers of a Dutch vessel cruising around Cape Leeuwin in 1622. How Captain Harris arrived at this conclusion is not known.

Mr J D Ogilvy, a Fellow of the Linnean Society for the study of natural history, provided his opinion too. He said that most authors were inclined to believe that the dingo is not indigenous to Australia, but was brought here through some human instrumentality.

Sir Frederick McCoy had been a professor and museum director specialising in palaeontology and natural history before his death in 1899. Professor McCoy was a divisive figure in Australian natural history studies, and promulgated some very bizarre theories. The article quoted the Professor's view that the dingo is one of the most ancient of the indigenous mammals of the country, and that it had abounded before man himself appeared.

A scientific author, Joseph McCabe – also quoted – claimed that after the Chalk Period (or Cretaceous Period, 146 to 66 million years ago), the ocean levels sank, and the upheaval drained the shallow sea north of Australia, forming a great bridge across the waters of Asia. Along this great bridge the animal and plant populations passed slowly into Australia.

No one had much of an idea in 1916, that much is apparent. It would take more than 200 years before the dingo's history was finally sorted out, thanks to genetic science.

Arthur Crocker perhaps pushed his luck a little too far by asking Robert Kaleski for his opinion. Kaleski, normally never short of an implausible theory, responded: 'The dingo's origin is wrapped in mystery; some fiercely asserting that he is

indigenous, others that the Dutch dropped him when they fled from the Gulf of Carpentaria. None of the beggars know. So, it's no use arguing about it.'

That evasive and meek non-answer was surprising, given that just two years earlier he had had no hesitation in giving his opinion in *Australian Barkers and Biters*.[3] Buoyed by Professor McCoy's 'findings', he had declared that, after a lifetime of studying canines, he believed the Australian dingo was the world's primal dog. It was derived from the nasty-looking, long-extinct marsupial lion through the Tasmanian tiger. He claimed that the dingo was the progenitor of all the canines, and that the felines sprang from the same source.

He went on to pooh-pooh the popular theory that the dingo came from Asia with travellers. That idea was untenable, he wrote, because everyone knew that man had only existed on the earth for about 500,000 years, but the dingo lived in Australia 3 million years before man even appeared. That, he said, effectively put paid to the imported-by-man theory.[4] So there!

Yet Kaleski tweaked his dingo-origin theory a tad for another newspaper article in 1916. This time, he got as close to being right as he ever would: 'After studying the subject for half a lifetime, I am convinced that the dingo was brought here either by the Chinese or Malays, in the form of the big red dog of China, probably a couple of hundred years ago.'[5] In his two estimations on the arrival of the dingo in Australia, Kaleski's calculations varied by a mere 2,999,800 years.

In 1936 the far-fetched propaganda continued with the following article that appeared in the *Sydney Mail*. No surprises there. The 'anonymous' author insisted that dingoes practised colour segregation and that certain colours made some dingoes more aggressive than others:

Just recently, in the Kilcoy district of Queensland, a female was caught with two pups, one black, the other brown. The reds, usually with a dark muzzle, are the worst sheep destroyers. They mutilate and cripple until they are exhausted, while the blacks and tans are milder. Again, it has been found that, in packs, the reds and yellows run together, whilst the blacks and tans do not intermingle.[6]

Kaleski – sorry, the anonymous author – then described the dingo bitch as the canine world's femme fatale. He (or she) told readers that a female dingo would attack and kill a female domesticated bitch on sight. Evidently a cunning, libidinous creature, she would then entice the bitch's guileless domestic mate away. After having her wicked way with him, she'd slay him too, the poor fellow. A deadly homewrecker, clearly out-crossed with a praying mantis.

The foregoing is typical of the quality of dingo-related information offered to urban Australians. The constant theme regarding the dingo's status throughout the late nineteenth and early twentieth centuries was a complete lack of balance and accuracy. The man everyone looked to for that information was completely out of his depth, and just made it up as the whim took him. It is little wonder many Australians have had but the haziest ideas of the dingo's natural history.

Through Kaleski's writings they knew the dingo had contributed to the cattle dog, but wild dogs were an unusual concept for Britons. They came from a tradition in which wolves, extinct from their island for several centuries, were merely figures of myths and fairy tales. And their portrayal was always negative.

Fly strike killed many more sheep than the dingo ever would, but the fly has never made very interesting reading. The dingo's depredations on Australia's flock were so immense that it is only natural they got bad press. People believe what they read and hear, and the dingo found itself friendless in its own land.

It was probably the lack of big cats or large, old-world wolves that led people to attribute the depredations of wild dogs to tigers and panthers living in the bush. There had to be something lurking in the bush that could terrify people, so people made up stories, or at the very least exaggerated. A lot.

* * *

Country people have always found great sport of making their city cousins' flesh creep with tales of the lurking terrors of the bush. Since colonisation there have always been reports of mythical creatures that haunted Australia's bushland.

The deadliest terrestrial creatures in Australia were spiders and snakes. The most dangerous large land mammal, until the coming of the Asiatic buffalo to northern Australia as tropics-hardy milking and meat stock in the early to mid-nineteenth century, was the kangaroo. The truth is that the wild dingo – as distinct from the habituated dingo – is no danger to people, but kangaroo bucks protecting a harem sometimes are.

A huge dingo was the most plausible means of creating an implausible bush monster. Dingoes already had an appalling reputation for bloodthirstiness, and it follows that if they were particularly huge they would also be nastier and prey on people. Nightly dingo howling wouldn't have helped. It is the eerie sound of rapacious wolves that plays on the primeval fears of man and terrifies a people who have never learned to understand the dingo or comfortably share the bush with it.

In the early twentieth century, stories began to circulate of monstrous wolf-like dogs that preyed on domestic stock. Suddenly, the bush was full of monster dingoes, but funnily enough, no evidence of any of the bush monsters was ever produced.

Reports of the killing of unknown wild animals in Gippsland, Victoria, and the Lachlan district and Liverpool Range in New South Wales, reached the national press in 1921.

The Gippsland case claimed that a very large animal, described as being about 3 feet 6 inches in height, with large claws and tusks, had killed about 1000 sheep in a few years.[7] It met its grisly demise through poisoning. The Lachlan monster, even more ferocious, went down swinging against a man armed with a rifle and assisted by three or four powerful dogs. And once again, it was powerful, and possessed of a large, square head and long teeth.[8]

When Robert Kaleski heard about (or created) these bush monsters, he naturally decided to share another hair-raising incident with his awestruck readership. It involved two stout-hearted fellows he claimed were in his casual employ, and through sheer, inexplicable coincidence it occurred in scrubland along the Heathcote Road, just a few miles from his farm:

> This district [Holsworthy] has at various times suffered
> from scares as to tigers, etc., being out in the back gullies.
> Between Bulli and Liverpool there lies some of the
> roughest, wildest, and most forbidding country on the

Australian continent – country which has rarely, if ever, been trodden by man. In this region two gum-getters who used to work for me – two fine bushmen – went out shooting one day near the head of the Woronora, and they came back at mid-day in a terrible hurry, half-scared out of their lives. They explained that in a gully in the main gorge they had heard a peculiar cry. On making an investigation they saw what they described as a huge cat-like creature, quite as large as a full-grown tiger. It was playing about on a flat rock in front of a small cave.

When the animal saw them, it gave an ugly snarl and disappeared into the cave. One man was armed with a Winchester .32 carbine, and the other a double-barrelled shotgun, and both were dead shots; but they were too frightened to shoot, and cleared off back to the camp at once. From what I know of them, I thought the devil himself couldn't have scared them; one, in particular, is more like a native in the bush than a white man. If I had had time, I would have organised a party to look for the brute, but was too busy.[9]

Nothing more was heard of the terrifying Woronora big cat. Perhaps it wisely decided to keep a low profile. Mythical bush-monster stories have frightened, amused and bemused Australians ever since. The problem was that Kaleski was serious, unlike the equally silly and fabled drop bear, created as a joke that is even perpetuated by the Australian Museum website:

Bush walkers have been known to be 'dropped on' by drop bears, resulting in injury including mainly lacerations and

occasionally bites. Most attacks are considered accidental
and there are no reports of incidents being fatal.

There are some suggested folk remedies that are said to
act as a repellent to Drop Bears, these include having forks
in the hair or Vegemite or toothpaste spread behind the
ears. There is no evidence to suggest that any such
repellents work.[10]

All the bush monsters including the now non-deadly drop bear
(we are all relieved to hear that) were outdone by another
mythical beast that was allegedly shot and killed close to
Sydney. Through another incredible coincidence, the monster's
calf-killing rampage ended not more than eight miles from
Robert Kaleski's farm. Our man in Liverpool, as earnest and
straight-faced as ever, takes up the story:

> There was another scare here at present on account of some
> wolves killing the calves in this country. One was shot
> recently on its beat by one of our best bushmen, F. Childs,
> of Macquarie Fields. He sent word he had got it, and could
> I come and identify it, as it had him puzzled. Hoping I had
> struck my missing link – the cross connecting the dingo
> with the Tasmanian Wolf [thylacine] – I rode over to see it.
> He had kept the skull and the skin. From these I judged it
> had been a large animal and though very like a dingo in
> general appearance it had two peculiarities. One of these
> was the tail, which had been very short (almost 'bobbed'),
> and curled over the back, after the manner of the Eskimo
> wolf-dogs. It was also unusually broad and bushy. On the
> underside of the body the hair was very long, and distinctly
> unlike that of the dingo, each hair being separate, instead

of close together. It had the usual dingo blackish stripe
down the back and tail. On each side of its mouth is a
tooth of abnormal length and quite tusk-like in shape. If
the dog had been some years older the probability is that
these tusks would have been much longer.[11]

Kaleski declared the Macquarie Fields monster the progeny of
an Eskimo wolf-dog and a dingo, and not much more than a
year old. He claimed that sled dogs, the remnants of Sir Ernest
Shackleton's Trans-Antarctic expedition, were taken to Mount
Kosciuszko (of all places) when their work in Antarctica was
done. They escaped, Kaleski claimed, and subsequently
crossbred with dingoes to create a huge, stock-killing monster.

That was another of Kaleski's big fat lies. Shackleton's dogs
were all killed and eaten by his starving expeditionary party
on floe ice in Antarctica in 1915.

Sensationalism continued to follow the dingo. Its
bloodthirsty reputation ensured that the tall stories kept on
coming, and newspapers published tiresome nonsense like this
prime piece published, unsurprisingly, in the *Sydney Mail* on
21 May 1928. The *Sydney Mail* was a weekly magazine version
of the *Sydney Morning Herald*. It featured photography, a large
sporting section, bold advertisements, interesting news items,
and it also published a lot of fiction. As we have seen.

In this instance, the Briagolong tiger had been terrorising
the Bairnsdale region of eastern Victoria. The creature had
destroyed £200 worth of sheep and created a minor reign of
terror. It was believed to be an animal escaped from a circus.

Then it had been caught and proven to be a huge dingo. In
death it measured 8 feet from nose to tail, and its general
appearance suggested to some observers that it was a new

species. It was reputedly vividly yellow, its tail large, bushy and white-tufted; down its back ran a vivid white stripe. It shared some very similar features with the mythical bush monster whose skin and skull Robert Kaleski had viewed at Macquarie Fields. Funny, that.

It was Sydney's *The Sun* that brought the next polar-bear-sized monster instalment into the lounge rooms of urban Australians. (It should be noted that Killabakh is not in western New South Wales; it is north of Taree, near the mid-north coast of the State.) Rendering the story in anything but its original form would do the nameless author with the Kaleski touch the gravest injustice:

At Killabakh, in the Western District of N.S.W., there is a dingo said to have a body as large as that of a polar bear that is terrorising the district, and is asserted to be the leader of the packs of dingoes which have invaded the mountains. In the latter days this dingo has killed six fair-sized calves and four pigs, and has dined on their carcases. Numbers of wild dogs are frequently seen in broad daylight sneaking around quite close to holdings and homes of settlers.

Until recently on Saturday afternoons and Sundays women and children spent their leisure hours in the bush and on the hills without the slightest fear of molestation from wild dogs. However, a change has come. Men walk warily while following their daily occupations in the bush alone and most people keep to their own firesides at night. Many reputable members of the Killabakh community have seen this canine monster. Two residents saw him capture and pull down a fair-sized wallaby. When they

> shouted out to frighten him he just trotted off carrying the
> wallaby in his mouth as though it were a rabbit. Parties
> will be organised to hunt him down.[12]

It appears that a posse was not able to be organised to run the
monster to ground. It was a common, and very disappointing,
theme in Kaleski's bush-monster stories.

But the bush-monster bug wouldn't stop biting any time
soon. On 25 May 1933, the *Cobram Courier* in Victoria reported
great jubilation among the farmers of the besieged Swanpool
and Mansfield districts. The cause of celebration was that a
monstrous dingo that had caused the deaths of hundreds of
sheep in the area had met a grisly end.

So destructive and fearsome had the animal become that a
reward had been offered for its destruction.

> Two intrepid fellows had started out on the monster's trail
> with guns and traps. They had set their traps, and coming
> back the next morning had found the animal sitting up on
> a trap, securely caught. It was extremely ferocious, and
> immediately showed fight, but soon received a charge of
> shot for its trouble. Local farmers said it was the largest
> dingo they have ever seen, and when weighed it turned the
> scales at 100 pounds.

Yeah, right.

The bush-monster myth cropped up at the most unexpected
times. The horrors of the Second World War were briefly
forgotten by a story in the *Inverell Times* of 2 October 1942.
Another man-eater was on the prowl in northern New South
Wales. It was initially thought to be a huge dingo, but then its

tracks were seen on soft, bare ground and were reckoned to be much too big for a dog. Later, the animal was allegedly seen. It stood about 3 feet high, and had a head as big as a bucket. What size of bucket was not divulged. It was said to be as big as a lion, and unsurprisingly, it eluded the lead-footed, muddle-headed incompetents hunting it.

The bush-monster stories are an amusing look back at the past, but they demonstrate how easy it was for the dingo to be reconfigured as the terror of the bush – a monster with a taste for terror and human flesh. The tall stories of giant, sheep-slaughtering dingoes were a progression from over a century of dingo bad press.

But the little knee-high dingo couldn't quite cut it alone as a monster. First they had tried to inflate the dingo's dimensions, but no matter how much imaginary size they pumped into it, you were only ever in danger from a dingo if you were a chook or a sheep. Or already dead.

To make a real monster out of the dingo, an exotic and dangerous admixture was needed. And Robert Kaleski found it happily trotting around the Sydney show rings with its devoted handlers. It was the German shepherd dog, known at that time in German-phobic Australia as the 'Alsatian wolf-dog'.

And as we are about to see, Kaleski was astonishingly successful in convincing susceptible graziers and the nation's gullible, trembling lawmakers that a creature that existed only in his mind threatened to destroy Australia's wool industry.

CHAPTER 16

The 'Alsatian Wolf-dog Menace'

As we've seen, after emerging from the trauma of the Great War, Australians weren't very fussed on Germany.

Robert Kaleski may have had his own reasons to dislike Germans. Poland, his father's homeland, suffered great devastation during the war when Germany invaded and fought Russia along the Eastern Front, which spanned Poland from north to south. The experiences of Kaleski senior and his Polish family may have influenced Robert's views.

In a remarkable coincidence, many of the Germans interned during the war were incarcerated at Holsworthy Army Camp, which neighboured Kaleski's property, Thorn Hill. Having those 'enemy aliens' next door may have also coloured his attitudes, because Robert Kaleski would often make it plain that he had no good opinion of Germany or its working sheepdogs. He ignored the German collie and utterly detested the Alsatian.

Kaleski's lack of interest in conventional breed origins and his obsession with superficial similarities rendered him susceptible to hasty, ill-considered conclusions. He laid eyes on the unlucky Alsatian, and from that point on it was the spawn of the wolf. If it was ever crossed with the dingo, he reasoned –

as only he could – it would certainly destroy the entire wool industry. Kaput!

The truth is there were hardly any German shepherds in Australia at the time, and all of them show exhibits, but the breed would be banished from Australia if Robert Kaleski had his way. The Alsatian wolf-dog gave him the ideal outlet for exercising his imagination and testing his national influence. Let's not forget that Kaleski claimed that huskies and even foxes hybridised with the dingo. The Alsatian was just an extension of his obsession.

The Australian public was extraordinarily naïve when it came to both dogs and Robert Kaleski. It was a dangerous mix. He could make all the claims he wanted, and had to prove nothing, because everyone took him for his word. Australia's great dog authority was beyond reproach.

Kaleski played on the fears of the most vulnerable – the wool-growers. They were copping a battering from the dingo, and the thought that the problem might be exacerbated was enough to put them in a state of high dudgeon. It is doubtful whether Kaleski ever achieved anything as influential as his Alsatian fear and smear campaign.

He got the ball rolling in 1928. Not long ago, he said, he attended a dog show in Sydney, and there he first saw the Alsatian. He was asked his opinion of the breed, and said he thought they might be useful as pets or watch dogs, and preferable as a novelty to some of the lapdog breeds, but he was concerned for the pastoralist.

It was a notorious fact, he said, that in the past Australia's pests were all introduced by well-meaning people, who honestly believed their efforts would help their country. Yet if the people who introduced the rabbit, the fox and the prickly pear were alive and could be identified they would probably be 'torn to pieces' by an enraged populace.

He feared that the Alsatian might prove to be as great a pest. If they escaped from the stations where they were kept as pets – none were being kept on sheep stations – then crossed with the dingo they would create a ferocious, cunning destroyer, twice as big and twice as troublesome as the dingo. (As was his wont, he reasoned that a large dog crossed with a medium-sized dingo would create a dog twice the size of the large dog, not a dog the same size as either parent, or somewhere in between both.)

Kaleski insisted (unconvincingly) that he had no quarrel with the Alsatian as a pet dog in towns. He was even generous enough to admit that some of them showed remarkable intelligence and tractability in America and Germany (but apparently not in Australia). He reminded his readers of his fictional story of Shackleton's huskies that had escaped confinement at Mount Kosciuszko. He had asserted that they crossbred with the dingo and the offspring made their way to the killing fields of Macquarie Fields on Sydney's outskirts, there to run rampant slaughtering and devouring cattle.

He ended his account wistfully, regretting the weighty burden of being a visionary: 'Prophesying future danger is a thankless job, but it must be done in view of the possibility of seriously damaging our world's monopoly in merino wool.'[1]

* * *

The German shepherd dog is a refined version of the old German shepherding dogs. Herr von Stephanitz created it at the end of the nineteenth century. It was usually called the

Alsatian in English-speaking countries, but the Kennel Club of Great Britain, whose decision-makers were ardent disciples of the shallow heresy of similarities, decided it had been bred directly from wolves and named it the 'Alsatian wolf-dog'. They used the same faulty similarities principle with the kelb tal-fenek (the Maltese rabbit hound), in naming it the pharaoh hound, because it looked vaguely similar to the jackal god Anubis of the pharaohs' tombs.

It is true that the German shepherd bears some vague similarity to the wolf. They both have erect ears and some share sable colouring. But there the similarities begin and end. Huskies and malamutes appear far more wolf-like than the German shepherd dog. Just as the kelpie needs to look dingo-like to survive in Australia, so the husky and malamute must look wolf-like to survive in the Arctic. The suggestion that German shepherds are bred from wolves is as incorrect as it is ridiculous.

Kaleski apparently found himself an influential ally, in the person of a New South Wales police detective named Lawrence who had been sent to Britain on an extradition matter. While in Lanarkshire, Scotland, he allegedly got involved in the hunt for a killer, to wit, an Alsatian.

Lawrence and Kaleski were obviously known to each other, and were equals in terms of dog IQ and propriety. Kaleski was the founder and Honorary Secretary of the Cattle and Sheepdog Club of New South Wales. Detective Constable Lawrence, fresh from his bloodcurdling fact-finding mission, was invited to Kaleski's club to talk about 'the Alsatian menace'.

Lawrence had come back from Britain with hearsay, nothing more, but he had a strong supporter in Kaleski, who had long been passionately in love with hearsay. It was all calculated to add credibility to and garner public support for Kaleski's agitation for a permanent prohibition on the importation of German shepherds.

The *Sydney Mail* published an article about Detective Constable Lawrence's concerns. There was no by-line, but you didn't have to be a detective constable to work out who had written it.

It said that Detective Constable Lawrence, of the Criminal Investigation Department, Sydney, had just returned from an official visit to Great Britain. He was decidedly on the side of 'our' organised graziers in 'their' desire for an embargo on the importation of Alsatian dogs. While Mr Lawrence had no profile in the Australian dog world, the article declared he was a 'well-known dog authority', and from what he had seen and heard in Britain he was convinced that there was a distinct danger that the Alsatian would develop into a big pastoral pest in Australia unless the government banned importations.

Lawrence was said to be struck by the resemblance between the Alsatian and the dingo, and referred to reports stating that the Alsatian circled sheep in the same way as the dingo did. Fancy that! He intended to bring the matter before the dog clubs. The article closed by stating that the graziers were hopeful that the Federal Government would not delay any longer in imposing the embargo.[2]

At the same time as Detective Constable Lawrence was denigrating a foreign sheepdog to the members of Kaleski's Cattle and Sheepdog Club of New South Wales, an eminent Scottish show judge, Mr T W Hogarth, arrived in Australia. He was very familiar with the Alsatian, and he had a vastly different take on the breed.

Mr Hogarth had seen Alsatians working in sheepdog trials in Germany, as well as in police-work displays and other exercises. In Western Australia, he met Mr Le Souef, the Director of the Perth Zoo, who told him that the dingo was no longer pure because of the domestic dog influence on it. (That was partly correct.) Mr Hogarth said there was therefore an admixture already running wild in the bush. He added that there can be little danger should the Alsatian go bush, as it was no more dangerous than any other breed of dog. The Alsatian, he said, was no more wolf-like than Pomeranians, Pekingeses, bloodhounds, terriers and all other breeds – which also have the wolf as their common progenitor.[3]

No one who mattered was very interested in Mr Hogarth's sensible opinion. 'Alsatians Banned' was the common headline in syndicated newspapers throughout Australia in the last week of May 1929.[4] The graziers' associations mobilised influential lobby groups and pressured the Federal Government. Kaleski's fear campaign had the Chicken Littles in Canberra slapping an ill-advised interim ban on the importation of Alsatians before the sky fell in.

The 'Alsatians Banned' article, in Kaleski's hallmark writing style, confirmed the five-year importation ban, but expressed disappointment that no action was being taken against Alsatians already in Australia. The Minister for Health, Sir Neville Howse, heroically declined to indicate the reason for the decision, which he said the government had arrived at after 'full consideration'.

The graziers' case was that 'the crossbreeding of Alsatians with dingoes had formed a breed whose depredations on sheep were reaching alarming proportions'. That was an outrageous lie. There were no proven cases where Alsatians had crossbred with dingoes. It was known, the article stated, that Alsatians had a wolf strain that intensified their inclinations: a theory that was vigorously denied by owners and importers of Alsatians. The Alsatian people argued that other large dogs such as greyhounds constituted an equal potential menace, and they were right.

The Alsatian people had every right to be up in arms: at the same time as the ban was placed on the Alsatians, feral kangaroo dogs were causing havoc in pockets around the country. But that issue was conveniently ignored. However, the panic-stricken pastoralists, wearied from the dingo wars and goaded by Kaleski into fearing worse, insisted that the ban on Alsatians was urgently required, and important to pastoral interests. They carried the day.

The government arrived at the decision after 'full consideration' – of what? Robert Kaleski's contrived fear campaign was all, because there wasn't a scrap of evidence against the Alsatian. And there never would be.

* * *

It is true that the German shepherd, like the Australian cattle dog, is not everyone's cup of tea. It has a positive attitude to its guarding duties and does not suffer fools gladly. Being twice the size of a cattle dog, it is an imposing, forceful dog when the occasion calls for it, and it has suffered the injustice of fashion, and the consequent high rate of incompetent ownership that inflicts itself on all the fashionable breeds. But back then German shepherds were a rarity in Australia, and there was absolutely no reasonable excuse for the vilification campaign.

Did the German shepherd pose a threat to sheep? An individual dog certainly might, but no more of a threat than any other breed. There were plenty of dingo-cross wild dogs and town dogs killing sheep; cattle dogs and cattle-dog crosses were regular offenders. Sheep-killing fox terriers (the classic 18 pound foxie) are known of. Kaleski, of course, had no idea what he was talking about.

If there was ever a blue-ribbon sheep killer it is the dingo, and it was making a fair fist of decimating the merino mobs without any assistance. But with Alsatians now the hot topic, it didn't take long before the dingo x Alsatian sheep-killer reports started pouring in.

After allegedly killing more than 200 sheep on Dyanberin Station in northern New South Wales, three huge dingoes (they were always huge), two pure-breds and one supposed dingo x Alsatian, were alleged to have been shot by landowners. One skin brought to Armidale measured 6 feet long, which sounds enormous to the reader but is a standard length for a dingo hide including the tail: 4 feet for the head and body, 2 feet for the tail. The landowners were alleged to have 'closely examined the half-bred dog and they were satisfied that it was the result of a cross between a dingo and an Alsatian'.[5]

There were hundreds of such articles appearing in syndicated newspapers all around the country around the same time. The common theme in every one of these reports was a complete lack of photographic or verifiable evidence, even though many articles claimed that the monsters were caught and skinned.

Not everyone got caught up in the hysteria. The level heads usually belonged to people who knew something about the breed, or dogs. In an address to a gathering of the Alsatian Defence League Club of Victoria, Mr Norman Mitchell a well-known barrister and the proprietor of one of the best-known Alsatian kennels in Victoria, outlined the history of the Alsatian breed. He stated the breed was founded in antiquity, but its modern form was the result of selective breeding out of the north and south German sheepdogs in the late nineteenth century, to obtain the best possible specimens. Since that date, he said, the Germans had kept a stud book for the dogs, and at the present time nearly half a million dogs were registered in Germany alone. The English and Australian dogs were bred from this German sheepdog, which was in fact that country's recognised national sheepdog, just as the kelpie was Australia's. (Though a German shepherd has not been born that could hold a candle to the kelpie as a sheep worker.)

Mr Mitchell said:

> it was nonsense to suggest that the Alsatian is any more
> predisposed to kill sheep than any other breed of dog, and
> the contention that he is harmful or a danger to sheep or
> human life is so much hot air. Substantial evidence is
> available now, as it was four years ago when Dr Robertson,
> the Commonwealth Director of Veterinary Hygiene,
> recommended against the ban on Alsatians and insisted
> that the contentions of the Graziers Association and others
> have no substance at all.[6]

In an address to the Geelong and District Kennel Club, Mr Mitchell said that in England, where there were 6 million more sheep than in Victoria, and about 100 times as many

Alsatians, the dogs were not regarded at all as being a menace to the industry, much less to human life.

In addition, he said that the Alsatian had been used in police work, Red Cross work during the First World War, and in leading the blind. Six hundred and fifty dogs had been donated by the German breed society at the close of the war to be used in leading blind soldiers. This aspect of the work had been taken up in America, Switzerland and England, and a movement was afoot in Victoria to import a trainer who would equip Alsatians already in Australia to do similar work here.[7]

Yet the voices of reason were few compared with the mass of articles, which snowballed as the years of the interim ban rolled on. There were literally hundreds of anti-Alsatian articles in virtually all Australian newspapers of the day. The hysteria grew and grew.

There was much argument – wrote Kaleski in the second (1933) edition of *Australian Barkers and Biters* – as to whether the Alsatian was part-wolf, and whether he was dangerous. 'Breeders of Alsatians contend that there is no wolf blood in them. They contend that the dogs' stud book in Germany goes back to 1891, and that they can definitely trace the ancestry of the Alsatian Wolf-Dog to the native Sheepdogs of Thuringia and Wurtemberg.'[8]

The German shepherd supporters actually knew what they were talking about, while Kaleski peddled damaging fantasy. The result of typical Germanic thoroughness, the German shepherd's development was a carefully recorded and managed strategy under Captain von Stephanitz's direction. The German shepherd is one breed that does *not* have a murky origin.

Conversely, Kaleski was not even accurate in tracing the origins of dog breeds developed on his doorstep during his lifetime. He had no idea of the Alsatian's ancestry, but the ridiculous name 'wolf-dog' was guilt enough for him. His personal and very successful vilification campaign was based on his obsession with similarities, his transparent exploitation of the prevalent anti-German sentiment, blatant fear-mongering, his practice of converting fantasy and hearsay into 'fact', and his most damaging weapon: his 'opinion'.

In the second edition of *Australian Barkers and Biters* Kaleski stated that his opinion of Alsatians had not changed since he first saw them at a show in Sydney twenty-five years earlier. He made the bizarre claim that he had been offered the use of them to cross with his cattle dogs. Why pedigree Alsatian owners would offer him the use of their dogs is anyone's guess, but Kaleski said he had emphatically refused on account of the 'wolf showing in them'. He harked back to his 1926 article in the *Sydney Mail*, where he 'drew attention to the danger to the pastoral industry of allowing the Alsatian to run wild and cross with the Dingo creating a cunning, ferocious destroyer which would give far more trouble than our dingo'.

He added that the Western Australian Canine Association had endorsed a report that had found the Alsatian wolf-dog should be prohibited from entering Australia, on the grounds '[of] its long list of convictions in Great Britain as a ruthless sheep-killer, that its intelligence did not exceed the cunning of the average crossbred dog, and that there was no useful canine function for the Alsatian [a sheepdog] in Australia'.

The 'long list of convictions in Great Britain as a ruthless sheep-killer' that Kaleski used as the cornerstone of his argument was nothing more than Detective Constable Lawrence's tale of the Lanarkshire Alsatian sheep-killer, contrived to fuel Kaleski's case. There has never been any evidence that the Lanarkshire 'attack' took place other than Lawrence's say-so. The odour of collusion hung thick over the detective and Kaleski, but Lawrence got away with his story because perjury is not an offence in a kangaroo court.

Kaleski continued: 'Western Australia became alive to the Alsatian danger and made very strong representations to the Commonwealth Government, with the result that the interim ban on the importation of Alsatians for five years became effective from June 1929.' He said that the Western Australian Government, caring more for its graziers' interests than any of the other States, then passed its own legislation against Alsatians, the *Alsatian Dog Act* of 1929, which required the sterilisation of all Alsatians in the State. Failure to comply would lead to very heavy penalties.

Let us be clear about Western Australia's cause to legislate against the Alsatian. The sole reason for that State's lily-livered, knee-jerk reaction was a statutory declaration from a dingo trapper named Arnold Herbert. Mr Herbert asserted that he had trapped and killed a male Alsatian dog that had escaped the care and control of a Mr Eric McManus of Mount Marshall. He further stated that he later trapped a dingo bitch with two crossbred Alsatian pups. He never claimed that the Alsatian or its offspring had been involved in sheep killing. One Alsatian does not a crisis make. A facsimile of that singular document is contained in *Australian Barkers and Biters*.

Kaleski went on to warn that the danger of Alsatians that crossbred with dingoes would create such a problem that 'all stock would have to be permanently guarded and that boundary riders, stockmen and drovers will have to work in armed numbers or else they would be killed and eaten'.

Kaleski then went on the full offensive, telling his gaping audience that the German police used their German shepherd police dogs as instruments of terror. 'The German police preferred a dog capable of serious attack. Their law, unlike Australian justice, does not presuppose innocence until guilt is proved.' With the hide of Jessie the Elephant, he accused German law of doing what he himself had done to the Alsatian: presuming guilt without a speck of evidence.

'In Germany, there is nothing like our sensitiveness to public opinion with regard to people being bitten. The German police-dog is a powerful animal capable of great severity. One sees persons being constantly attacked in Germany by such dogs in a manner that would never be tolerated over here.'

Kaleski could not have done more to vilify a breed of dog, not to mention Germans and German coppers. He added a lame, insincere declaration that he had nothing at all against the breed. It wouldn't have placated or convinced Alsatian owners, or anyone else with anything between their ears.

* * *

The Commonwealth Government called for and received a report by Dr Robertson, the Director of the Commonwealth Division of Veterinary Hygiene. His report clearly and emphatically stated that 'any dog might attack man or sheep since there are savage representatives in every breed, but I can see no reason why the Alsatian should be singled out as being either vicious or a sheep killer'.[9]

The jittery Federal Coalition Government, terrified of inciting the wrath of the rural sector, ignored Dr Robertson's report. Flaky sitting members preferred hearsay and alarmist theory to professional advice. The Kaleski-fuelled clamour and clangour of the graziers' associations received a sympathetic hearing in Canberra. Commonwealth legislation was passed prohibiting the importation of Alsatian dogs into Australia:

> *I, Sir Isaac Alfred Isaacs, the Governor General, acting with the advice of the Federal Executive Council, do hereby prohibit the importation into the Commonwealth of Australia of Alsatian Dogs unless the consent in writing of the Minister of State for Trade and Customs has first been obtained.*
>
> *This proclamation may be cited as Customs Proclamation No. 265. Given under my Hand and the Seal of the Commonwealth this sixth day of June, in the year of our Lord, 1934 and in the 25th year of His Majesty's reign.*
>
> *By His Excellency's Command, THOMAS W. WHITE Minister of State for Trade and Customs.*
>
> *God save the King*

Well may they have said 'God save the King', because nothing would save the Alsatian.

Thirty-three years earlier, in December 1901, the *Immigration Restriction Act* had been enacted. It had been among the first pieces of legislation introduced to the newly formed Federal Parliament. It had been designed to limit non-British migration to Australia, and allow for the deportation of 'undesirable' people who had settled in any Australian colony prior to Federation. We know it as the White Australia Policy, and it beat the Alsatian Customs Proclamation No. 265 of 1934 by a short half-nose as the most bigoted, ill-advised piece of legislation ever enacted by Australia's Federal Government.

One would have thought Robert Kaleski would be crowing. He had got his way: the Alsatian had been banned from being further imported into Australia, almost entirely due to his opinion that it was a danger to the wool industry.

But Kaleski was disillusioned with the Federal Government. He was disappointed that the Commonwealth had not ordered every last Alsatian in Australia destroyed. But the Federal Government had washed its hands of the actual dirty work of legislating for the compulsory euthanasing of extant Alsatians and left it to the States to manage their own affairs in that hugely unpopular regard. It would have been a brave State government that ordered people to start killing their pets based on no evidence.

* * *

Kaleski may have been taking some well-deserved heat from the Alsatian exhibitors and other rightfully concerned people because his next newspaper article appeared in the *Sydney Mail*, under the transparently pointless pseudonym 'Drover'.

'The position with regard to the Alsatian in Australia is truly Gilbertian,' he sniffed, 'but there was nothing funny about it. He said Western Australia had taken action to protect its flocks and herds, but so far, no legislation had been passed in the other States, and the measures proposed to be submitted to the New South Wales Parliament that required Alsatians to be desexed if in a grazing area were totally inadequate.

The truth was that the sheep flocks in Australia would have been at no greater risk from the Alsatian if every government had just sat on their hands. The Federal Government had decided to take the path of least electoral resistance. The graziers were a powerful and influential lobby group, and the managers of Australia's most important primary industry. It was easier to cheese off a few dog owners than risk the consequences of not acceding to Big Wool's wishes.

'Drover' claimed no one knew how many Alsatians there were in Australia. The German Shepherd Club did. There were less than sixty at the time of the ban. He claimed their numbers had increased 'alarmingly' ever since the federal interim ban was announced in 1929, but he had no way of knowing that. He then spuriously claimed there were comparatively few towns in either New South Wales or Queensland where pure-breds or 'half-breds' were not to be seen.

Robert Kaleski possessed a sense of infallibility the Pope would have killed for. And he was never troubled by the apprehension of spreading untruths. He claimed that it was indisputable that the Alsatian had already crossbred with the dingo. His only proof of that was his phony newspaper articles in the complicit *Sydney Mail*.[10]

It was certainly an impassioned plea from Australia's great dog authority, who failed to see the great contradiction in his argument. The creature he insisted would be wool's great destroyer was a cross between a dingo and a sheepdog. The dog he cherished and championed above all others, the Australian cattle dog, was, as he repeatedly (and incorrectly) told the nation, the product of matings between the dingo and a sheepdog. The difference between the two, the difference that engendered such axe-grinding vehemence in Robert Kaleski was one word: German.

* * *

'Drover' was up to his old tricks again in the *Sydney Mail* a few months later, this time ramping up the Alsatian menace as an imminent threat to human safety: 'Something drastic will have to be done before the menace goes too far to control. History will repeat itself for certain unless the imported menace is eradicated. Once an Alsatian-dingo crossbreed becomes established in the country, not only sheep and cattle, but human life, will be endangered.'[11]

To bolster his claims Kaleski presented another concocted story. He claimed, without a shred of evidence, that on not one but two separate occasions, Alsatian dogs had been 'lost' between Mount Isa and Camooweal in northwestern Queensland. That area, he reminded the reader, was inhabited by thousands of dingoes. Then he asked the Alsatian-lovers if they would be bold enough to say that mating with the dingoes would not take place, because he had proof that it had, and that things had taken a potentially deadly turn for the worse:

> An Alsatian and a half-bred dog have been raiding goat
> yards at night in Mt. Isa, and parents take every precaution
> to keep their young children in at nights. Some days ago, a
> butcher at Mt. Isa shot an Alsatian dog which had been
> worrying his sheep, and its mate was joined by another
> dog. It is a pity the butcher did not send the skin of the
> Alsatian to Canberra as an exhibit in the case.

Like every sensational Kaleski exposé, the conclusion to the Mount Isa story was a disappointing 'It is a pity'.

'Drover' went on to agitate for dire action to rid the country of the Alsatian menace. The graziers' associations, he warned, 'may soon be compelled to take the law into their own hands, and they would have the sympathy of the vast majority of people'. He called for immediate and drastic action because the danger was now too great. 'The imported killers must not be tolerated under any consideration and the State Governments must act immediately and not only sterilise, but deport or destroy all Alsatians.' And Kaleski was happy to drop the graziers in it. He wrote that the graziers would be happy to compensate all the Alsatian owners. He warned that if the menace was allowed to grow, 'war against Alsatians will be waged with a vengeance'. Tough talk is cheap.

A second contributor to the *Sydney Mail*, 'Mulga', continued the offensive with a new angle that was so shrill and inaccurate that it is not hard to conclude it was written by the Alsatian's great nemesis himself, or an equally ill-conditioned acolyte. He warned 'that thousands of Alsatians were to be found in all parts of the country and were being deliberately bred and sold to rabbiters and kangaroo hunters which increased the danger to the pastoral industry'.[12]

He admitted that 'so far, no definite evidence of this cross had been produced'. Oh really? But he said that a correspondent at Liverpool (Kaleski) had stated that 'the wild dogs that have recently been troublesome in that district appear to be a mixture of half-bred dingo and Alsatian wolf-dog'.

Displaying the typical Kaleski-esque trait of jumping to improbable conclusions, 'Mulga' pointed the finger at an advertisement that offered 'nine weeks old pups, wolf and stag hound cross with greyhound bitch', the father and mother both being described as 'champion kangaroo and fox killers'. Thus, Mulga tersely warned, the menace grew. He had of course concluded that 'wolf' meant Alsatian wolf-dog, not Irish

wolfhound. Off and running, with the bit between his teeth, Mulga forewarned doom and destruction, based on an innocuous advertisement for kangaroo dogs.

Here, said Mulga, was someone deliberately breeding a mongrel killer while the authorities were powerless to intervene. The 'wolf' cross only had to breed with the dingo to 'produce a horde of hunting marauders capable of spreading the utmost havoc throughout the country'. Yeah, right.

* * *

Legislation was passed in New South Wales in 1933 that required all Alsatians to be desexed in certain Pasture Protection Board areas. In 1939 in Western Australia, section 2 of the *Alsatian Dog Act 1929*, was amended to make it unlawful for any person in Western Australia to have in their possession or control any Alsatian 'wolf-hound'.

Some agitators were lobbying for Alsatians to be shot on sight on rural land, and in 1934 the Northern Territory Government banned the importation of Alsatians and empowered police to shoot them on sight.

An Alsatian Protection League was formed in the affected States by desperate and disappointed Alsatian owners and sympathisers. They protested to their local, State and federal ministers, but their concerned voices of reason fell on deaf ears.

The question of whether Alsatian dogs were a menace was raised in the Queensland Parliament in 1935, when the Minister for Lands, Mr Pease, was asked if the government would consider further legislation against Alsatians. The minister said that he had seen references to Alsatians in the press, but whether or not the Alsatian, as distinct from other large-sized dogs, was a menace had not been definitely determined. He said the only official report on the issue was that supplied to the Commonwealth Government (and quoted earlier) by Dr Robertson, Director, Federal Division of Veterinary Hygiene.[13]

The following year, the Campaign Director of the Alsatian Shepherd Dog Defence League, Mr Marienthal, called out the Queensland Government or anyone else who claimed to have a dingo–Alsatian hide in their possession and offered to donate a cheque to Brisbane Hospital in return for it.

Mr Marienthal said government files were thick with examples of attacks by other breeds on sheep, which he had forwarded from testimonials and statements from many sources. Time after time, various shires sent skins to the government, allegedly from dingo–Alsatian cross-breeds. In every case, he said, the skins proved to be those of pure-bred dingoes. If it was decided that all dog breeds must be destroyed or removed from pastoral areas, the league must agree, he said. But if it was decided to single out the Alsatian, the league must emphatically protest.[14]

Brisbane Hospital never received the Alsatian Shepherd Dog Defence League's donation. Despite hundreds of claims that thousands of Alsatians were crossbreeding with dingoes, not a single specimen, dead or alive, was ever produced. After the predictable spike in alleged Alsatian–dingo depredations that always turned out to be dingo or generic feral dog attacks, if they occurred at all, the hysteria died down. No one who supported the ban ever seemed to think, or wanted to admit, they had been duped.

* * *

Other than in Western Australia and the Northern Territory, Robert Kaleski failed in his effort to convince the State (and Territory) governments to sterilise or destroy the Alsatians that remained in Australia. People could keep their Alsatians, but if they lived in a certain pasture protection board area in New South Wales, Victoria or Western Australia, their dogs had to be desexed.

Still, Alsatian owners and their supporters refused to let the matter go, and they tried, decade after decade, to have the ridiculous ban lifted.

Politically, there was little benefit in repealing the ban. Other than during the Second World War and for a few years afterwards, Australia was governed by conservative Liberal–Country Party coalitions who would do nothing to further agitate or provoke their pastoral constituents.

One of the few surviving examples of submissions to the Federal Government to have the importation ban rescinded was drafted by Tony Parsons in August 1955 and was titled 'The Case for the Alsatian or German Shepherd Dog for the Orange Alsatian Protection Association'.

Parsons's submission was a thorough and lengthy examination of virtually every detail regarding the development and worldwide distribution of the German shepherd dog. It is an embarrassment to Australia that its law-makers required such a document after twenty-one years of proof that the Alsatian posed no threat to the wool industry. The 1950s was a golden decade for wool in Australia, and Alsatians were the least of the wool-growers' problems. In fact, they were no problem at all.

Parsons spelled out the case for repealing the ban so succinctly that even Blind Freddie could see there was no justification for the Alsatian importation ban. He said it all in concluding his submission:

> All of the above only serves to illustrate that there is no case against the Alsatian at all. It is a loosely strung together fabrication bristling with inaccuracies and devoid of any attempt to find evidence. Anyone who agrees with the ban will join the company of men to go down in history as the group who outlawed one of the greatest breeds in the history of the civilised world.

The problem was that the Federal Minister for Agriculture at the time was Mr Roger Nott, not Blind Freddie. Though the minister highly commended the submission, saying it was the most complete analysis of the subject yet received, the arch-conservative Menzies Government was not to be moved.

The minister's praise was cold comfort for Tony Parsons and the Alsatian owners of the Orange region of New South Wales, whom he was assisting out of a sense of justice. The Bathurst Pasture Protection Board had been mulling over whether to order the sterilisation of all Alsatians within its jurisdiction, which included Orange.

All up, the importation ban remained for forty years. Australian German shepherd breeders were left with a tiny, overworked gene pool with which to maintain their breed. Overly close breeding occurred, and temperamentally undesirable specimens were produced. It was unavoidable, but no one who could have helped was interested in their problems.

* * *

The ill-informed and discriminatory way the Australian federal, State and Territory legislators managed the 'Alsatian wolf-dog fiasco' was a stark contrast to the attitude of the rest of the world. While Australia was treating the German shepherd dog like a mindless killing machine, and its owners like second-class citizens, in other nations it was the dog of choice for police, the military, and search and rescue organisations. It was the first seeing-eye dog, and won bravery awards, obedience awards, community service awards and agility awards. It starred in twenty-seven movies and a 166-episode television series (in which no sheep were harmed).

The German shepherd was considered the most talented dog in the world. It just took Australia forty-five years to work out it had been conned. But there's your federal leadership for you.

Saving the Australian wool industry from the imaginary menace of the Alsatian wolf-dog was Robert Kaleski's most notable achievement. He would release several editions of *Australian Barkers and Biters* and would write and publish *Dogs of the World* in 1947.[15] Its execrable subtitle, 'Showing the Origin of the Canine and Feline Species from the Australian Marsupial Lion About Three Million Years Ago', is enough to demonstrate that nothing had changed Kaleski's laughable biological theories since he first published *Australian Barkers and Biters* in 1914.

He died aged eighty-four, on 1 December 1961, the year *The Adventures of Rin Tin Tin* premiered in Australia. Rinnie was the most popular dog in the country, but banned nonetheless. And no one seemed to notice or mind that *The Adventures of Rin Tin Tin* was set fifty years before the German shepherd was even developed. Yo Rinnie!

Robert Kaleski's importation ban was still in place when he died. And would be for another unlucky thirteen years.

* * *

By the 1970s, lobbying increased pressure on the Australian Government. German shepherds were serving with distinction around the world, and it was patently and embarrassingly obvious, even to the graziers, that they posed no threat to pastoralism.

In 1972, after twenty-eight consecutive years of conservative government, the Whitlam-led Labor Government ushered in an era of social change in Australia. They weren't frightened to tread on conservative toes and were prepared to listen to reason. They agreed to a one-year importation trial of the German shepherd dog in 1973.

Of course, no issues arose during those twelve months, and Federal Customs Minister Lionel Murphy permanently lifted the import ban in 1974. After forty-five years of baseless discrimination, German shepherd enthusiasts were able to gain access to overseas bloodlines and enjoy the same rights as every other dog owner in Australia.

So ended the con job that was the Alsatian wolf-dog menace. Yet sheep graziers all over mainland Australia were still fending off attacks by the real wild dogs recruited to the dingoes' cause. And there wasn't an Alsatian wolf-dog to be seen among them.

CHAPTER 17

Tess and Zoe of the Thin Blue Line

I n the first half of the twentieth century, there was no law enforcement publication more austere or chauvinistic than the English *Police Gazette*. In 1940, sensational stories circulated in the world press about the incredible feats of a New South Wales Police dog, goading the *Gazette*'s bemused editor into not only questioning the validity of the stories, but also ridiculing them in his singularly jocular manner:

> On another page is reported the case of an Australian
> Police Dog that has been trained to take orders by radio.
> As the report comes from Australia and not from America
> 'the home of tall tales' we suppose there is something in it.
> Doubtless when the dog has been trained to track persons
> it will complete the job by escorting them to the Police
> Station, taking their names, filling in the Charge Sheet,
> and conducting the prosecution for the police, the vocal
> part being effected by barking in Morse.[1]

The radio-controlled police dog making all the news was Zoe. She was certainly the most famous dog in Australia, and other than Rin Tin Tin, the most famous dog in the world. Happy are those who have not seen yet still believe; the *Police Gazette's* sceptical editor, would have become scepticaller and scepticaller had he known Zoe's full repertoire, because taking remote orders via radio was just one of many strings to her multi-talented bow.

The emergent New South Wales Police Dog Unit of the 1930s did spectacularly well with what it had to work with, because the standard of Alsatian available in Australia during the years of the Alsatian importation ban left much to be desired. The impetus behind the incredible feats of Zoe and her predecessor, Tess, was arguably Australia's greatest ever dog man, Constable Adam (Scotty) Denholm, the godfather of the Australian police dog units and the Australian service dog industry in general.

* * *

Police dogs were nothing new in the early twentieth century, though in Australia they were latecomers to public service.

Since the days of the Sydney Cove penal colony, people have always found it easy to get themselves hopelessly lost in the surrounding bushland, and almost as easy to get themselves killed in it. In the early years it was absconding convicts, who usually perished from exposure or starvation. As Sydney grew and crept out to the edges of the surrounding wilderness, the lure of the bush caused people of all ages, but often children, to come to grief. Huge search parties were often unable to find the lost in time to save their lives.

By the early 1900s, overseas fact-finding missions had the New South Wales Police hierarchy thinking that dogs would be a big help in finding the lost and tracking felons. It was the lengthy and arduous pursuit of a double murderer that finally convinced them that police dogs should be trialled.

In 1932, William Cyril Moxley murdered a young couple near Liverpool and bamboozled the police when he made off into the bush. They followed an endless number of false trails and dead ends in all directions before Moxley was eventually captured in the scrub behind Narrabeen on the northern beaches, more through good luck than good management.[2]

Bill McKay, the Metropolitan Superintendent, was a border-collie fancier, and he reckoned that police tracker dogs would have run Moxley to ground in short order. He was determined to develop a dog unit run along the same lines as those of the British, European, North American and South African police forces: a general patrolling and ready-response tracking unit.

In 1932 he set up Australia's first police dog kennels on wasteland at Alexandria, in inner Sydney, and hoped to recruit dog handlers from within the force. What he did not know was that the future heart and soul of his dog unit was one of his current boys in blue, walking city beats just a few blocks from police headquarters.

* * *

Adam Denholm was the son of Scottish emigrants, who had sailed for Australia in 1910 when he was just five years old. At school in Goodna in rural Queensland, his broad native brogue earned him the nickname 'Scotty', a name that followed him for life.

The Denholm men were bona-fide dog men. Scotty's father, Davy, kept highly trained collies and lurchers, as *his* father had done before him. Scotty's first dog was a kelpie bitch named Gyp, and he was just six years old when he brought her home in his school shirt as a three-week-old pup.[3] Over the years, Scotty and his father taught her all manner of routines and tricks. Before he left Goodna to seek his fortune in Sydney, Scotty served a long dog-training apprenticeship with his father. He could not have received better instruction in dog management.

By the time he became a general-duties police constable, stationed at The Rocks in Sydney in the early 1930s Scotty Denholm was mad about Alsatians. For an all-round utility dog they were hard to beat, despite the nonsense that had led to their Australian importation ban.

His first Alsatian was a bitch, whom he named Gyp after his kelpie. He and Gyp were inseparable, and he proudly took her home to Goodna to visit his father while on annual leave. His father, dog-less at the time, thought Scotty had brought Gyp as a gift for him. Scotty did not have the heart to disappoint his father, so he left her behind when he returned to Sydney, determined to get himself another Alsatian.

He found a jet-black bitch he called Darkey. Off duty, he was a regular sight around Centennial Park in eastern Sydney, training Darkey to do over a hundred tricks and activities.

It was while he was on leave that Scotty learned from a dog friend that the New South Wales Police were setting up a dog unit and were about to advertise for trainers.[4] He wasted little time in attending police headquarters, seeking an opportunity to audition with Darkey. He and another constable named Reilly were given a chance to demonstrate their ability to McKay and Police Commissioner Walter Childs.

The auditions were held at Willoughby in the northern suburbs in early 1932, close to the veterinary practice of

Charles Court-Rice. Court-Rice was a senior figure in Sydney's dog world, and a respected dog judge. It was he who had rewritten a breed standard as opposed to Kaleski's for the Australian cattle dog in 1911. It is likely Superintendent McKay sought his advice on establishing the dog unit.

The Sydney press were also at the audition, and the next day the newspapers were glowing in their descriptions of Denholm's and Reilly's demonstrations. The New South Wales Dog Unit was soon up and running.

Superintendent McKay had found himself two very capable trainers, but the unit would start with no dogs. Denholm had previously agreed to sell Darkey to a Mr Hill of Newcastle in order to help finance his forthcoming wedding, and Reilly's dog, a Belgium-bred Alsatian with the posh name, Haras de St Géry, was too old for police work.

Bloodhounds were the tracking dog of choice in Europe, North America and South Africa. The English police also used bloodhounds for tracking work and Airedale terriers for patrolling. McKay had envisaged bloodhounds as his dog unit's canine recruits, but the dogs were just about more trouble than they were worth. They are stubborn, slow-thinking, wilful dogs that have their share of health issues. Denholm had a very poor opinion of their worth as an all-round police dog, and questioned their ability to perform in the extremes of the Australian bush. The other obstacle to using bloodhounds was that there were none to be had at that time in Australia.

The other breed Denholm considered was the Dobermann. Again, there were none in Australia at the time, though the Austrian and South African police used them and were enamoured with their all-round ability. Approaches were made to the South African police to secure a pair for the New South

Wales Dog Unit, but the South Africans, while agreeing to provide dogs, insisted the Australian handlers be sent to them for some months of formal training. That wasn't going to happen, Superintendent McKay was probably pushing his luck just getting the little dog unit funded, and the Dobermann option was off the table.[5]

So, McKay's dog unit started training the only dogs readily available in Australia that were suitable for the purpose: Alsatians. Denholm and Reilly were more than happy with that. Several puppies were donated to the dog unit, and of course, the breed's noisy, high-profile detractors raised the predictable clamour.

The Alsatian's courage and determination could not be questioned, and crossbreeding with dingoes and slaughtering sheep on Denholm's watch was highly improbable. So, grasping at straws, they claimed that the Alsatian would be useless for tracking duties. Denholm, a man ahead of his time, knew that the agile, problem-solving Alsatian, which had reasonable tracking ability, was a better police dog prospect than the stubborn, slow-thinking, super-scenting bloodhound, and events would prove he was right.

Banned from importation the reviled Alsatian might have been, but under the expert tutelage of Denholm and Reilly, the Alsatians of the New South Wales Police Dog Unit were soon rubbing their detractors' noses in their unfounded criticisms.

* * *

There were two standout performers in the new dog unit. One was Reilly's black and cream male dog, Harada, named after a visiting Japanese tennis player; the other was a black and tan bitch, Olympian Bonanza, Denholm's pride and joy. He renamed her Tess.

Elementary training for the Alsatian puppies began with developing scenting ability through hide-and-seek games. Once the puppies were accustomed to using scent as their main form of detection, they were deemed suitable for police work training, particularly search and rescue.

Obedience training commenced immediately after the puppy had proven its scenting worth. The inbreeding in Australia caused by the importation ban, coupled with other poor breeding practices, had seen a rise in the timidity of many Alsatians. To counter any inherent nervousness, the puppies were taken out into the wider community and were incrementally exposed to every conceivable stimulus.[6] Much had been written about training Alsatians, but Denholm ignored the books and followed his instincts when training his police dogs. He was rarely wrong, and taught his charges an incredible repertoire of commands and routines.

In the dark days of the Great Depression, when Sydney lawmen were battling large organised crime gangs like the vicious razor gangs of the inner city, the newly formed dog unit was the force's greatest public-relations asset. The press reported on the unit's every activity, and Tess and Harada became household names. Reilly and Harada left the dog unit soon after both dogs began operational work.

Tess's first foray into search and rescue work involved locating a lost rabbit hunter in the bushland gorges of French's Forest. The man, a confident bushman, had ventured too far away from his vehicle and found himself stranded as night fell. His worried family contacted the police and Denholm took Tess out to help search for him.

The man's coat was given to Tess to scent and, surging ahead in the pitch-black night, she wasted no time in finding the man asleep and unharmed under a rock overhang, waking him with a vigorous face-licking. She then nudged him to his feet, and using her body as a guide she walked him to where Denholm waited above.

The *Sydney Morning Herald* reported the story, quoting the rescued man as saying, 'Tess was marvellous. I can't speak too highly of the dog, and the police. It was remarkable that I should have been found in that dense scrub in pitch darkness. The dog seemed almost human, and was a credit to the trainer.'[7]

The *Sydney Morning Herald* ran another Alsatian story that same day. It reported that an amendment to the Pastures Protection Bill had been introduced to the New South Wales Legislative Assembly, requiring anyone who possessed an Alsatian dog within certain pastures protection districts to sterilise it.[8] The two stories gave a fair indication of the gulf between the heroics of Scotty Denholm's dog unit and the hysterical self-interest attacking the Alsatian in Australia at the time.

Denholm's dogs would go on to prove their worth time and time again in the rugged bushland around Sydney and throughout regional New South Wales. The stories of Tess's sagacity are too numerous to relate. Yet the greatest impediment to her success was the often-unacceptable delay in requesting assistance from the dog unit. The passage of time or wet weather obliterated scent trails and the subject of the search was often found too late. But on a discernible trail, Tess was never wrong, no matter how wrong she appeared to be at the time. Denholm quickly learned never to second-guess her.

Such was Tess's skill that on a sad Christmas Day in 1937 she not only located the body of a murdered little girl midstream in a creek, but she also immediately led Denholm to the murderer, who lived nearby.[9] That remarkable feat shot her to national prominence in newspapers around Australia. Robert Kaleski and the Alsatian-haters had nothing to say.

If there were ever a nation that appreciated a courageous working dog, it was Australia. The heroics of working cattle dogs and sheepdogs had paved the way for Denholm's police dogs. For every rescued child, for every captured felon – and there were many of both – the public's reverence for the dogs of the New South Wales Police Force grew.

Tess and Scotty Denholm were now household names, and when not on active duty the dog unit found itself another niche in entertaining the public. The popularity of these police-dog exhibitions had everything to do with Denholm's outstanding showmanship. Leaping into moving vehicles, apprehending 'robbers' and disarming pistol-wielding 'felons' were all just part of the show.

Tess was a genuine star, but as the years rolled by, Denholm realised he needed to find an understudy for her. As things turned out, it was from an unexpected acquisition that Denholm was able to breed his next star.

* * *

In 1935 an unnamed woman attended the Alexandria dog unit yard and asked to speak to Scotty Denholm. She wished to make a donation of an aged and also unnamed Alsatian bitch that she could no longer accommodate.

Checking the pedigree, Denholm thought that the old bitch's bloodlines were a good match for Kaspar, a male dog of the unit. Because of the Alsatian importation ban the Australian gene pool was tiny, and care needed to be taken to ensure mated dogs were not too closely related. Denholm took her and mated her with Kaspar.

There were three puppies in the litter: a black dog pup and two white bitches, one evidently better than the other. The dog pup went to the bitch's previous owner, and Denholm and another dog unit member Constable Spicer flipped a coin for the two bitches.

Spicer ended up with the better specimen and he called her Zoie. Denholm named his pup Blondie. Midway through Blondie's elementary puppy training, he was heartbroken to discover that she was stone-deaf. With no effective means to control an attack-trained dog, Denholm reluctantly decided that Blondie should be euthanased.

Not long afterwards, Constable Spicer was transferred to a rural station, and due to the Alsatian wolf-dog fiasco he could not consider taking Zoie with him. Denholm bought her for £5 and changed the spelling of her name to Zoe.[10]

Zoe became an accomplished actor and entertainer. Her incredible ability, her showy white coat and short, snappy name ensured that everyone remembered her. But it wasn't until Tess died in 1942 that Zoe really came into her own as a performer. She amazed audiences at agricultural shows, exhibitions, police carnivals and gymkhanas and in theatres all over New South Wales, and was the only dog ever to perform at the Sydney Town Hall.

In her heyday, she knew over 300 routines; there seemed to be nothing she couldn't learn.[11] She even appeared in several

At one time, Zoe was the most famous dog in the world. She is pictured here with the radio receiver that allowed her to be handled remotely by Denholm. (Courtesy of State Library of New South Wales, IE1298376)

Cinesound and Fox Movietone short films and two full-length Cinesound productions. Director Ken Hall called her 'One Take Zoe', such was her reliability before the camera.[12]

With such an impressive CV, it was no surprise that Zoe was also an outstanding police dog. Denholm considered her to be the best tracker he ever trained and when one considers the skills of her canine colleagues, it is high praise indeed. She was responsible for tracking down and apprehending several dangerous felons, and her problem-solving ability and courage knew no limits. Australia adored Zoe.

Zoe even had her own miniature car, which she was able to drive so competently under remote instruction that she actually passed a driving test at Sydney's Domain and was issued with a driver's licence. A miniature army tank with a coin collection well was made for her to raise funds for the Red Cross during the Second World War, and as unbelievable as it sounds she was even taught to fly her own miniature plane. For safety's sake, it remained attached to a main pivot point, but she flew the little aircraft around in circles with such competence that Denholm was fully convinced that Zoe would be capable of flying it unattached.[13]

Denholm collaborated with the Tasma Radio Company in 1940 to design a radio receiver for her to wear on a harness.[14] The date was set for Zoe's first radio-controlled routine, but parts for the receiver were delayed in arriving from Holland. In the end, Denholm had just four short days to train Zoe to respond to remote handling for the scheduled exhibition.

The press, radio technicians and police officials all attended to see Zoe put on a spectacular show. She completed an arduous obstacle course, climbed ladders, walked tightropes, jumped through hoops and carried billy cans full of water, which she scooped from a tub after turning a tap on and off. On command, she discarded her collar and put it back on again. All the while, Scotty Denholm sat high in the grandstand and gave her directions through a microphone and transmitter.[15]

It was that startling exhibition that made headlines all around the world. Denholm wanted to use the remote device in police operational work, but with war raging against Japan, all radio transmitters were called in.

By war's end, Zoe was old, deaf and retired from active police work and entertaining.

* * *

Unhappy in service, Dot the police dingo looks pretty unimpressed at performing for Denholm. All Dot's activities were conducted on-lead. (*The Cold Nose of the Law*, C Bede Maxwell, 1948)

At the same time as Zoe was developing her police dog training skills, Denholm took delivery of a bloodhound male named Disraeli, and shortened his name to Dizzy. He was an excellent tracker, though prone to wanting to follow a scent trail over cliffs when people had jumped or fallen. On several occasions it took all of Denholm's strength to prevent Dizzy, and himself, from plunging to their deaths.[16]

Dizzy was incapable of learning anything other than tracking, and despite the bloodhound's famous scenting skills, his tracking ability never exceeded that of Denholm's well-trained Alsatians. He was a one-off trial that convinced Denholm and his colleagues that they had made the right choice in the Alsatians.

The other unique Denholm experiment was Dot the police dingo. Taronga Zoo in Sydney presented him with a pair of dingo puppies. One died through misadventure in the dog yard. But the little bitch, Dot, was able to be taught a series of basic obedience commands, though experience taught Denholm to keep her on a lead at all times. Dot was easily able to scale 8-foot-high fences with ease, a feat that none of the much larger Alsatians could manage.[17]

Typical of any tamed dingo, Dot became more and more distant and intractable as she matured. She killed three dogs in the yard, including a fox terrier puppy that Denholm was training as a gift for his son. Eventually he returned her to Taronga Zoo, where her training earned her a role in the zoo's mini-circus.[18]

* * *

Scotty Denholm, Tess and Zoe were Australia's police- and service-dog pioneers, and the inspiration for generations of service-dog handlers. Never have the dogs of any Australian service arm enjoyed a higher or broader profile. Australia has since produced plenty of great service dogs and handlers, but considering the times and the obstacles Denholm and his colleagues faced, none who came after them ever produced dogs that exceeded Tess's or Zoe's remarkable achievements.

Denholm served in the dog unit for fourteen years. Following his resignation from the New South Wales Police Force, he was sought after by the entertainment industry as an animal trainer and consultant, and trained all the animals for the long-running *Skippy* television series.

The service-dog industry provides security for millions of Australians every day. Police dogs provide a huge range of services, from search and rescue operations to drug and

explosives detection, crowd control and screening, the apprehension of offenders and high-risk tactical operations. In New South Wales alone, five dogs have lost their lives in the line of duty, such is the dangerous nature of their work.

Australia owes much to the six-year-old boy from Goodna who in 1911 came home with a much-too-young kelpie puppy tucked inside his school shirt.

Conclusion

For every dingo that has run down sheep, there've been a couple of sheepdogs working them. In paddocks or in yards, backing sheep (running over their backs) in races, penning them up in shearing sheds, the working dog has given Australia's wool industry the impetus to survive the dingo's assault of a thousand cuts.

Beef has suffered less at the hands of the dingo. The cattle industry's own dogs, the heelers, tailed or stumpy-tailed, worked harder cattle on large holdings in more remote regions. Sometimes the ideal cattle-working dog was a compromise cross between the kelpie and the heeler. Some progressive cattle men worked their cattle with pure kelpies.

Whatever the industry and whatever the breed, the common thread to Australian pastoralism was the home-grown working dog.

The Australian working dogs were created during the age of colonial transportation and hard labour, then the opening up of the interior to wool and beef, and they played as important a role in the forming of the nation as any group of human workers. The working-dog heroics that put a Sunday roast in the oven, balls of wool in the haberdashery, a suit on a rack, or a steak on the plate were brought to Australia's cities through word of mouth, newspaper accounts or bush poems

and stories, but nothing does a sheep or cattle dog more justice than seeing it work first-hand.

In an age when internet technology has reached almost every corner of rural Australia, the working dog's role is unchanged, and will remain so. The Australian working dog is the embodiment of the original role of the dog – the super-tool that facilitated agrarianism and a settled, prosperous society.

Never has a workforce produced so much yet asked for so little. Not that the koolie, the heelers or the greatest of them all, the kelpie, has ever complained. No grievances about the boss or the food, and no strikes for better pay or conditions. The Australian working dog, whether a rural stock worker or an urban police dog, has always been much too busy just getting on with the job.

Australia owes an enormous debt to its incredible working dogs, and the men and women who created and worked them. Until now, this has not extended to official recognition of their vital role in Australia's development. History is written by the victors, and the victors have forgotten their dogs.

It's about time the dogs that made Australia received their due.

Endnotes

Chapter 1: The Dingo Conquers Prehistoric Australia

1   J-F Pang, C Kluetsch, X-J Zou et al, 'mtDNA data indicate a single
     origin for dogs south of Yangtze River, less than 16,300 years ago,
     from numerous wolves', *Molecular Biology and Evolution*, Vol. 26,
     Issue 12, 1 December 2009, pp 2849–2864, academic.oup.com/
     mbe/article/26/12/2849/1536110.

2   Tim Flannery, *The Future Eaters: An Ecological History of the
     Australasian Lands and People*, Reed Books, Melbourne, 1994,
     pp 166–169.

3   A Ardalan, M Oskarsson, C Natanaelsson, A N Wilton, A Ahmadian
     and P Savolainen, 'Narrow genetic basis for the Australian dingo
     confirmed through analysis of paternal ancestry', *Genetica*, Vol. 140,
     Issues 1–3, March 2012, pp 65–73.

4   Flannery, op cit, pp 170–173.

5   Ibid, p 170.

6   Bradley Smith, *The Dingo Debate: Origins, Behaviour and
     Conservation*, CSIRO Publishing, 2015.

7   Ibid, p 85.

Chapter 2: The British and the Dingo Get Acquainted

1   William Dampier, *A Voyage to New Holland: The English Voyage of
     Discovery to the South Seas in 1699*, ed James Spencer, Alan Sutton,
     Gloucester, 1981.

2   James Cook, *Captain Cook's Journal During His First Voyage
     Round the World Made in HM Bark* Endeavour, *1768–71*, ed
     W J L Wharton, Elliot Stock, London, 1893, pp 242–248.

3   Joseph Banks, *The* Endeavour *Journal of Sir Joseph Banks: 1768–1771*, ed J C Beaglehole, Public Library of New South Wales in association with Angus & Robertson, 1962.

4   Cook, op cit, p 279.

5   Johannes Caius, *Of Englishe Dogges: The Diversities, the Names, the Natures and the Properties*, trans Abraham Fleming, Bradley, London, 1880.

6   Ralph Clark, *The Journal and Letters of Lt. Ralph Clark 1787–1792*, ed Paul G Fidlon and R J Ryan, University of Sydney Library, purl.library.usyd.edu.au/setis/id/clajour, journal entry of 2 August 1787.

7   Ibid.

8   Arthur Phillip et al, *The Voyage of Governor Phillip to Botany Bay*, John Stockdale, London, 1789.

9   Watkin Tench, *A Narrative of the Expedition to Botany Bay*, Debrett, London, and Chamberlaine, Wilson, White, Byrne, Gruebier, Jones and Dornin, Dublin, 1789.

10   John White, *Journal of a Voyage to New South Wales*, Debrett, London, 1790.

11   John White, letter to *Public Advertiser* (UK), 31 December 1790.

12   White, *Journal of a Voyage to New South Wales*, entry of 21 July 1788.

13   'Collection 10: George Bouchier Worgan – letter written to his brother Richard Worgan, 12–18 June 1788. Includes journal fragment kept by George on a voyage to New South Wales with the First Fleet on board HMS *Sirius*, 20 January 1788 – 11 July 1788', State Library of New South Wales; first published as *Journal of a First Fleet Surgeon*, Library Council of New South Wales in association with the Library of Australian History, Sydney, 1978.

14   John Hunter, *An Historical Journal of the Transactions at Port Jackson and Norfolk Island*, Stockdale, London, 1792.

15   Tench, op cit.

16   David Collins, *An Account of the English Colony in New South Wales*, Cadell & Davies, London, 1802.

17 White, *Journal of a Voyage to New South Wales*, entry of
    30 February.

18 Ibid.

19 Ibid.

20 Richard Johnson, letter to Evan Nepean. State Library of
    New South Wales, http://archival.sl.nsw.gov.au/Details/
    archive/110331129#. The Reverend Richard Johnson was
    appointed as the first chaplain to the colony of New South Wales
    in 1787, an appointment he held until 1800 when he returned
    with his family and Governor John Hunter on HMS *Buffalo*.

21 Collins, op cit.

22 Collins, op cit.

23 Phillip et al, op cit.

24 Sydenham Edwards, *Cynographia Britannica*, C Whittingham,
    London, 1800.

Chapter 3: Colonial Hounds Save the Day

1   Tench, op cit.

2   Ibid.

3   Collins, op cit.

4   Tench, op cit.

5   Collins, op cit.

6   Ibid.

7   Phillip, op cit.

8   Ibid.

9   Robert Hughes, *The Fatal Shore*, Collins Harvill, London, 1987.

10 Tench, op cit.

11 George Barrington, *The History of New South Wales*, M Jones,
    London, 1802.

12 Tench, op cit.

13 Barrington, op cit.

14 Ibid.

15 Banks, op cit.

16   Hunter, op cit.

17   White, op cit.

18   Hughes, op cit, pp 145–147.

19   Tench, op cit.

20   Hugh Dalziel, *British Dogs: Their Varieties, History, Characteristics, Breeding, Management and Exhibition*, Bazaar Office, London, 1879, pp 41–42.

21   Ibid.

Chapter 4: Sydney's Dogs Behave Badly

1   Elizabeth Macarthur Onslow, *Early Records of the Macarthurs of Camden*, Angus & Robertson, Sydney, 1914.

2   Ibid.

3   Ibid.

4   Collins, op cit.

5   See Australia, Parliament, Joint Library Committee and Frederick Watson, 'Proclamation of Governor Philip Gidley King, Tuesday, 17 February, 1801', *Historical Records of Australia: Series 1 – Governors' Despatches to and from England*, Vol. 3, 1801–1802 (Library Committee of the Commonwealth Parliament, 1915) p 50.

6   Hughes, op cit.

7   *Sydney Gazette and New South Wales Advertiser*, 7 June 1807.

8   Collins, op cit.

Chapter 5: Three Dogs Conquer Van Diemen's Land

1   David Hunt, *True Girt: The Unauthorised History of Australia Volume 2*, Black Inc, 2016.

2   David Owen, *Thylacine: The Tragic Tale of the Tasmanian Tiger*, Allen & Unwin, Sydney, 2003.

3   Barrington, op cit.

4   Harris, G. P. (George Prideaux) & Hamilton-Arnold, Barbara, *Letters and papers of G.P. Harris, 1803–1812: Deputy Surveyor-General of New South Wales at Sullivan Bay, Port Phillip, and Hobart Town, Van Diemen's Land*, Arden Press, Sorrento, Vic, 1994.

5   Hughes, op cit, p 228.

6   Barrington, op cit.

7   Ibid.

8   John West, *The History of Tasmania*, Vols 1 and 2, Henry Dowling, Launceston, 1852.

9   Barrington, op cit.

10   Ibid.

11   James Backhouse Walker, *Early Tasmania*, John Vail, Launceston, 1902.

12   Ibid.

13   'Eaglehawk Neck: History', Parks & Wildlife Service Tasmania, http://www.parks.tas.gov.au/index.aspx?base=2589.

14   Caius, op cit.

15   Thomas Bewick, *A General History of Quadrupeds*, S Hodgson, R Beilby & T Bewick, Newcastle Upon Tyne, 1790.

16   Henry Melville, *The History of the Island of Van Diemen's Land from the Year 1824 to 1835 Inclusive: To Which is Added, a Few Words on Prison Discipline*, Smith & Elder, London, 1835.

17   Charles White, *Early Australian History: Convict Life in New South Wales and Van Diemen's Land, Parts I and II – The Story of the Ten Governors and the Story of the Convicts*, C & G S White, 'Free Press Office', Bathurst, 1889.

18   Parks & Wildlife Service Tasmania, op cit.

19   Walter Beilby, *The Dog in Australasia*, George Robertson & Co, Melbourne and Sydney, 1897.

Chapter 6: The Kangaroo Dog, the Dingo and the Birth of Wool

1   Hughes, op cit, p 634.

2   Richard Hough, *Captain Bligh and Mr Christian*, Hutchinson & Co, London, 1972, p 290.

3   Ibid.

4   Hughes, op cit, pp 299–300.

5   Letter from Nicolas Baudin to Philip Gidley King, 23 December
       1802, F M Baden (ed), *Historical Records of New South Wales, Vol 5:
       King, 1803, 1804, 1805*, Government Printer, Sydney, 1897,
       p 830.

6   'Batman's Treaty', State Library of Victoria, ergo.slv.vic.gov.au/
       explore-history/colonial-melbourne/pioneers/batmans-treaty.

7   Hans Kruuk, 'Surplus killing by carnivores', *Journal of Zoology*,
       Vol. 166, Issue 2,
       February 1972, pp 233–244.

8   Charles St Julian and Edward K Silvester, *The Productions, Industry,
       and Resources of New South Wales*, J Moore, Sydney, 1853.

9   National Museum Australia website, Defining Moments in
       Australian History, Rabbits Introduced. http://www.nma.
       gov.au/online_features/defining_moments/featured/rabbits_
       introduced.

10   'Huge wild dog captured', *The Chronicle*, Adelaide, 22 August 1935,
       p 4.

11   'Wild kangaroo dogs: Boy menaced', *Western Star and Roma
       Advertiser*, Toowoomba, 25 January 1933, p 3.

12   'Kangaroo dog menace', *National Advocate*, Bathurst, 26 August
       1937, p 4

Chapter 7: The Beardies and Bobtails Find a Home

1   Iris Combe, *Herding Dogs*, Faber & Faber, London, 1987, pp 14–15.

2   Ibid.

3   Ibid, pp 28–29.

4   Desmond Morris, *Dogs: The Ultimate Dictionary of Over 1,000 Dog
       Breeds*, Ebury Press, London, 2001, p 408.

5   John Jamieson, *An Etymological Dictionary of the Scottish Language*,
       Vols I and II, University Press, Edinburgh, 1808.

6   Noreen R Clark, *A Dog Called Blue: The Australian Cattle Dog
       and the Australian Stumpy Tail Cattle Dog 1840–2000*, WriteLight,
       Sydney, 2003, pp 149–150.

7   Bill Robertson, *Origins of the Australian Kelpie: Exposing the Myths and Fabrications from the Past*, Bill & Kerry Robertson, Ballan, Victoria, 2015.

8   Combe, op cit, p 52.

9   www.convictrecords.com.au/convicts/timmins/james/133817, State Archives of New South Wales; Series: NRS 1151.

Chapter 8: Hall's Heelers and Timmins Biters Build a Beef Empire

1   'George Hall', Australian Royalty, www.australianroyalty.net.au/individual.
php?pid=I43525&ged=purnellmccord.ged.

2   'The Old Great North Road', New South Wales Office of Environment & Heritage, www.environment.nsw.gov.au/nswcultureheritage/TheOldGreatNorthRoad.htm.

3   Clark, Noreen, op cit, p 9.

4   Ibid, p 9.

5   Bewick, op cit.

6   Ibid.

7   Albert (Bert) Howard, *The Timmins Story: Australian Origins and Heritage Files*, self-published, 2012.

8   Val Close, *Corn and Pumpkins and Yarramundi: Family History of James Timmins and Ann Baldwin*, self-published, 1986.

Chapter 9: The Making of Australia's Collie

1   Tony Parsons, *The Kelpie*, Penguin, Melbourne, 2010.

2   Ibid, p 91.

3   Ibid, pp 77–85.

4   Barbara Cooper, 'The Forbes Sheep Dog Trial and Kings Kelpie', Working Kelpie Council, www.wkc.org.au/Historical-Trials/Forbes-Trial-&-King's-kelpie.php.

5   Parsons, op cit, p 263.

6   Ibid.

7   Ibid.

8   Cooper, op cit.

9   'Obituary: Mr John Quinn', *Cootamundra Daily Herald*, 19 January 1937, p 3.

10   Ibid, p 42.

11   Ibid, p 310.

12   Ibid, p 312.

Chapter 10: The Missing Piece of the Kelpie Puzzle

1   Parsons, op cit, pp 67–68.

2   Mary S. Ramsay, 'Forlong, Eliza (1784–1859)', Australian Dictionary of Biography, http://adb.anu.edu.au/biography/forlong-eliza-12921.

3   Combe, op cit, pp 101–105.

4   Combe, op cit, pp 100–104.

5   Robertson, op cit, p 91.

6   Ibid, p 175.

7   Morris, op cit, p 446.

8   'Bedouin Shepherd Dog', www. molosserdogs.com/m/articles/view/1319-bedouin-shepherd-dog.

9   Parsons, op cit, pp 42–45.

10   Bert Howard, 'The Kelpie Story', Australian Origin and Heritage Files.

11   Barbara Cooper, 'The Story of Jack Gleeson', Working Kelpie Council of Australia, http://www.wkc.org.au/About-Kelpies/The-Story-of-Jack-Gleeson.php.

12   Parsons, op cit, p 57.

13   William C H Parr et al, 'Cranial shape and the modularity of hybridization in dingoes and dogs: Hybridization does not spell the end for native morphology', *Evolutionary Biology*, Vol. 43, No. 2 (2016), pp 171–187.

14   Dan Wheelahan, 'Dingo skull resistant to change from cross breeding with dogs, research shows', University of New South Wales Newsroom, 10 March 2016, newsroom.unsw.edu.au/news/

science-tech/dingo-skull-resistant-change-cross-breeding-dogs-
research-shows.

15   Shannon Verhagen, '"Wild" dingo characteristics are
genetically dominant', *Australian Geographic*, 11 March 2016,
www.australiangeographic.com.au/news/2016/03/Dingo-genes-
dominant-over-domestic-dogs.

16   Parsons, op cit, p 64.

17   Ibid, p 63.

18   Ibid, p 63.

Chapter 11: Cattle Dog and Kelpie 'Myth-information'

1   Mary Gilmore, *Hound of the Road*, Angus & Robertson, Sydney,
1922.

2   Robert Kaleski, *Australian Barkers and Biters*, NSW Bookstall
Company, Sydney, 1914.

3   Ibid, Endeavour Press, Sydney, 1933 (2nd edition).

4   Ibid.

5   Ibid, p 270.

6   Parsons, op cit, p 268.

7   Clark, Noreen, *A Dog Called Blue*, self-published, Wallacia, NSW,
2003, p 18.

8   Kaleski, op cit, 1933.

9   Ibid.

10   *Australian Encyclopaedia*, Angus & Robertson, Sydney, 1925–1926
(1st edition), 1958 (2nd edition); Grolier Society of Australia,
Sydney, 1977 (3rd edition), 1983 (4th edition); Australian
Geographic, Sydney, 1988 (5th edition), 1996 (6th edition).

11   Clark, op cit, p 58.

12   Howard, op cit.

13   Clark, op cit, p 22.

14   Parsons, op cit, p 38.

15   Clark, op cit, p 24; Berenice Walters, letter to Garry Somerville,
obedience judge and handler, 1 May 1974.

Chapter 12: Cattle Dogs and Kelpies in the Show Ring and Suburbs

1   Kaleski, op cit.

2   Ibid, p 26.

3   Parsons, op cit, p 122.

4   Ibid, p 230.

5   Ibid.

6   Parsons, op cit, p 230.

7   Clark, op cit, p 124.

8   Parsons, op cit, p 239.

9   *Hammersley News*, 6 December 1979.

Chapter 13: The Trials of the German Collie

1   Max von Stephanitz, *Der Deutsche Schäferhund in Wort und Bild*,
     Vereins für Deutsche Schäferhunde, Augsburg, 1901.

2   Ibid,

3   Edwin Hodder, *The History of South Australia, from its Foundation
     to the Year of its Jubilee, with a Chronological Summary of all the
     Principal Events of Interest up to Date*, Sampson Low, Marston,
     London, 1893.

4   www.australian-koolies.info/about-koolies/history-of-the-koolie.

5   Beilby, op cit, p 302.

6   'Wartime internment camps in Australia', National Archives of
     Australia, www.naa.gov.au/collection/snapshots/internment-
     camps/introduction.aspx.

7   Gerhard Fischer, 'German experience in Australia during WW1
     damaged road to multiculturalism', *The Conversation*, 22 April
     2015, theconversation.com/german-experience-in-australia-
     during-ww1-damaged-road-to-multiculturalism-38594.

8   National Archives of Australia, op cit; Moya McFadzean,
     'Internment during World War II Australia', Museums Victoria
     Collections, 2008, collections.museumvictoria.com.au/
     articles/1618.

9   Fischer, op cit.

10   www.australian-koolies.info/about-koolies/koolie-fundamental-menu.

Chapter 14: The Wild Dog Wars

1   L-J Thompson, H Aslin, S Ecker, P Please and C Trestrail, *Social Impacts of Wild Dogs: A Review of Literature*, ABARES, Canberra, 2013, prepared for AWI Ltd, www.wool.com/globalassets/start/on-farm-research-and-development/sheep-health-welfare-and-productivity/pest-animals/wild-dogs-foxes-and-pigs/wp525_wild_dog_management_in_australia.pdf.

2   John Pickard, 'Predator-proof fences for biodiversity conservation: Some lessons from dingo barrier fences', Chris Dickman, Daniel Lunney and Shelley Burgin (eds), *Animals of Arid Australia: Out On Their Own?*, Royal Zoological Society of New South Wales, Sydney, 2007, pp 197–207.

3   L R Allen and E C Sparkes, 'The effect of dingo control on sheep and beef cattle in Queensland', *Journal of Applied Ecology*, Vol. 38 (2001), pp 76–87.

4   Ibid, p 78.

5   Renate Kreisfeld and James Harrison, AIHW National Injury Surveillance Unit, Research Centre for Injury Studies, Flinders University, South Australia, Dog Related Injuries, 2005.

6   *National Wild Dog Action Plan: Promoting and Supporting Community-Driven Action for Landscape-Scale Wild Dog Management*, WoolProducers Australia, 2014, www.pestsmart.org.au/wp-content/uploads/2014/09/NWDAP_FINAL_MAY14.pdf.

Chapter 15: Bush Myths and Monsters

1   Arthur Crocker, '"The dingo pup." Some questions answered. Origin of the "native dog."', *The World's News,* 15 January 1916, p 6.

2   Walter Kilroy Harris, *Outback in Australia, or, Three Australian Overlanders: Being an Account of the Longest Overlanding Journey Ever Attempted in Australia with a Single Horse, and Including Chapters on Various Phases of Outback Life*, Garden City Press, Letchworth, UK, 1913.

3   Kaleski, 1914, op cit.

4   Ibid.

5   Robert Kaleski, 'The dingo', *Sydney Stock and Station Journal*, 20 June 1916.

6   Robert Kaleski(?), 'Menace of the dingo in Queensland', *Sydney Mail*, 25 November 1936.

7   'Mysterious Animal Killed in Victorian Scrub', *The Kyogle Examiner*, NSW, 24 July 1934.

8   Unknown Wild Animals, *Newcastle Morning Herald and Miners' Advocate*, NSW, 18 October 1921.

9   Robert Kaleski, 'Strange wild animals', *The Land*, Sydney, 23 September 1921.

10   https://australianmuseum.net.au/drop-bear.

11   Robert Kaleski, 'Unknown Wild Animals', *Newcastle Morning Herald and Miners' Advocate*, 18 October 1921.

12   Robert Kaleski(?), 'Wild dogs invading north coast led by giant dingo "The King of Killabakh"', *The Sun*, Sydney, 3 June 1922.

Chapter 16: The 'Alsatian Wolf-dog Menace'

1   Robert Kaleski, 'The Alsatian wolf-dog: Will it become a menace to our sheep industry?', Sydney Mail, 21 May 1928.

2   Robert Kaleski(?), 'The Alsatian threat', *Sydney Mail*, 13 February 1929.

3   'Are Alsatians a menace? Scottish expert thinks there is little danger', *Register News-Pictorial*, 16 April 1929.

4   Robert Kaleski, 'Alsatians banned', *The Queenslander*, 30 May 1929.

5   'Destroyed 200 sheep. Three huge dingoes shot. One Alsatian–dingo cross?', *The Advocate*, Burnie, 4 July 1933.

6   Article, *The Argus*, Melbourne, Wed, 19 July 1933.

7   'Not a menace: Lecturer defends Alsatians', *Horsham Times*, Victoria,
    11 July 1933.

8   Kaleski, 1933, op cit.

9   'Alsatian dogs favourable official report', *The Land*, 6 December
    1929.

10 Robert Kaleski (Drover), 'Menace of the Alsatian', *Sydney Mail*,
    11 July 1934.

11 Robert Kaleski (Drover), 'The Alsatian menace extends', *Sydney
    Mail*, 5 September 1934.

12 Mulga, 'Still the Alsatian', Sydney Mail, 15 May 1935.

13 'Government not yet sure whether Alsatians are a menace',
    The Telegraph, Brisbane, 30 October 1935.

14 'Dingo–Alsatian Skin: Alsatian Defence League would like to see
    one', *The Courier-Mail*, Brisbane, 15 August 1936.

15 Robert Kaleski, *Dogs of the World: Showing the Origin of the Canine
    and Feline Species from the Australian Marsupial Lion about Three
    Million Years Ago, the Most Notable Dogs Springing from It, and
    Illustrations Therewith*, William Brooks & Co, Sydney, 1947.

Chapter 17: Tess and Zoe of the Thin Blue Line

1   C Bede Maxwell, *The Cold Nose of the Law*, Angus & Robertson,
    Sydney, 1948, p 89.

2   Ibid, p 13.

3   Ibid, p 1.

4   Ibid, p 14.

5   Ibid, p 18.

6   Ibid, p 25.

7   Ibid, p 30.

8   Ibid, p 33.

9   Ibid, p 51.

10 Ibid, p 71.

11 Ibid, p 75.

12 Ibid, p 81.

13 Ibid, p 77.

14 Ibid, p 85.

15 Ibid, p 86.

16 Ibid, p 137.

17 Ibid, p 163.

18 Ibid, p 167.

Acknowledgements

When American music legend James Taylor was asked by Kerry O'Brien to end a *7.30 Report* interview with a quick comment on his life's good fortune, he responded by saying, 'Gratitude is the right attitude, and I'll probably end with that platitude.'

Being wholly sensible of Mr Taylor's lively lyrical wit and laudable sentiment, I am inspired to reflect upon my own good fortune and end with an expression of my gratitude for the people who made *The Dogs that Made Australia* possible.

Virginia Lloyd, my agent. Without Virginia's belief in me and my story, my book idea would have remained just that. Always the right advice, always the right direction, always in my best interests.

Vicki Hull, my sister and border collie woman: generous with her sounding board, her valued opinion, and her moral and material support. Our shared love of dogs and the bush-dog stories runs back to our early childhood and Vicki has been a positive influence during the project.

Tony Parsons, OAM, Australia's Mr Kelpie, and one of our great dog men: author, novelist, journalist, and our direct link to the men and women of the early Australian working dog world. Tony has been unstinting in his invaluable advice and personal recollections, proofreading and provision of documents, books and rare photographs.

Bert Howard. Bertie has had a bit going on over the last few years, yet he welcomed me into his home, opened his

treasure chests of colonial gold, and happily gave us the world's greatest dog stories. Ninety years of age and nothing has ever been too much trouble.

Mary Rennie, Shannon Kelly, Emma Dowden and all the team at HarperCollins. Such a fruitful and enjoyable collaboration: the belief, the vision, the clever editing, the beautiful production, the brilliant cover *and* Ned Heeler's menacing glare. Perfect!